Mathematical Models of Plant-Herbivore Interactions

CHAPMAN & HALL/CRC
Mathematical and Computational Biology Series

Aims and scope:

This series aims to capture new developments and summarize what is known over the entire spectrum of mathematical and computational biology and medicine. It seeks to encourage the integration of mathematical, statistical, and computational methods into biology by publishing a broad range of textbooks, reference works, and handbooks. The titles included in the series are meant to appeal to students, researchers, and professionals in the mathematical, statistical and computational sciences, fundamental biology and bioengineering, as well as interdisciplinary researchers involved in the field. The inclusion of concrete examples and applications, and programming techniques and examples, is highly encouraged.

Series Editors

N. F. Britton
Department of Mathematical Sciences
University of Bath

Xihong Lin
Department of Biostatistics
Harvard University

Nicola Mulder
University of Cape Town
South Africa

Maria Victoria Schneider
European Bioinformatics Institute

Mona Singh
Department of Computer Science
Princeton University

Anna Tramontano
Department of Physics
University of Rome La Sapienza

Proposals for the series should be submitted to one of the series editors above or directly to:
CRC Press, Taylor & Francis Group
3 Park Square, Milton Park
Abingdon, Oxfordshire OX14 4RN
UK

Published Titles

An Introduction to Systems Biology: Design Principles of Biological Circuits
Uri Alon

Glycome Informatics: Methods and Applications
Kiyoko F. Aoki-Kinoshita

Computational Systems Biology of Cancer
Emmanuel Barillot, Laurence Calzone, Philippe Hupé, Jean-Philippe Vert, and Andrei Zinovyev

Python for Bioinformatics, Second Edition
Sebastian Bassi

Quantitative Biology: From Molecular to Cellular Systems
Sebastian Bassi

Methods in Medical Informatics: Fundamentals of Healthcare Programming in Perl, Python, and Ruby
Jules J. Berman

Chromatin: Structure, Dynamics, Regulation
Ralf Blossey

Computational Biology: A Statistical Mechanics Perspective
Ralf Blossey

Game-Theoretical Models in Biology
Mark Broom and Jan Rychtář

Computational and Visualization Techniques for Structural Bioinformatics Using Chimera
Forbes J. Burkowski

Structural Bioinformatics: An Algorithmic Approach
Forbes J. Burkowski

Spatial Ecology
Stephen Cantrell, Chris Cosner, and Shigui Ruan

Cell Mechanics: From Single Scale-Based Models to Multiscale Modeling
Arnaud Chauvière, Luigi Preziosi, and Claude Verdier

Bayesian Phylogenetics: Methods, Algorithms, and Applications
Ming-Hui Chen, Lynn Kuo, and Paul O. Lewis

Statistical Methods for QTL Mapping
Zehua Chen

An Introduction to Physical Oncology: How Mechanistic Mathematical Modeling Can Improve Cancer Therapy Outcomes
Vittorio Cristini, Eugene J. Koay, and Zhihui Wang

Normal Mode Analysis: Theory and Applications to Biological and Chemical Systems
Qiang Cui and Ivet Bahar

Kinetic Modelling in Systems Biology
Oleg Demin and Igor Goryanin

Data Analysis Tools for DNA Microarrays
Sorin Draghici

Statistics and Data Analysis for Microarrays Using R and Bioconductor, Second Edition
Sorin Drăghici

Computational Neuroscience: A Comprehensive Approach
Jianfeng Feng

Mathematical Models of Plant-Herbivore Interactions
Zhilan Feng and Donald L. DeAngelis

Biological Sequence Analysis Using the SeqAn C++ Library
Andreas Gogol-Döring and Knut Reinert

Gene Expression Studies Using Affymetrix Microarrays
Hinrich Göhlmann and Willem Talloen

Handbook of Hidden Markov Models in Bioinformatics
Martin Gollery

Meta-analysis and Combining Information in Genetics and Genomics
Rudy Guerra and Darlene R. Goldstein

Differential Equations and Mathematical Biology, Second Edition
D.S. Jones, M.J. Plank, and B.D. Sleeman

Knowledge Discovery in Proteomics
Igor Jurisica and Dennis Wigle

Introduction to Proteins: Structure, Function, and Motion
Amit Kessel and Nir Ben-Tal

Published Titles (continued)

Chapman & Hall/CRC Mathematical and Computational Biology Series

Mathematical Models of Plant-Herbivore Interactions

Zhilan Feng

Donald L. DeAngelis

CRC Press
Taylor & Francis Group
Boca Raton   London   New York

CRC Press is an imprint of the
Taylor & Francis Group, an **informa** business

A CHAPMAN & HALL BOOK

CRC Press
Taylor & Francis Group
6000 Broken Sound Parkway NW, Suite 300
Boca Raton, FL 33487-2742

© 2018 by Taylor & Francis Group, LLC
CRC Press is an imprint of Taylor & Francis Group, an Informa business

No claim to original U.S. Government works

Printed on acid-free paper
Version Date: 20170719

International Standard Book Number-13: 978-1-4987-6917-4 (Hardback)

Visit the Taylor & Francis Web site at
http://www.taylorandfrancis.com

and the CRC Press Web site at
http://www.crcpress.com

I dedicate this book to my daughter Haiyun and my son Henry for all your love and support
  –Zhilan Feng

I thank my wife Lie for her love and patience
  –Donald L. DeAngelis

Contents

Foreword

In the latter half of the twentieth century, research directed toward obtaining a mechanistic understanding of the causes and effects of plant anti-herbivore defense became the focus of intense research in ecology and evolution. Part of this research effort has been the development of a diverse set of mathematical models of these mechanisms. The intent of this book is to introduce and summarize the current state of these modeling efforts. Professor Feng and Dr. DeAngelis have admirably achieved this.

This book begins with a sound introduction to basic mathematical theories of general predator-prey interactions such as the Rosenzweig–MacArthur equations. This introduction is followed by consideration of how these equations have been used to mathematically analyze interactions between plants and their herbivore predators. Then more recent mathematical models of plant herbivore interactions, such as linear programming models, are discussed, and in this introduction the notion of plant chemical anti-herbivore defense is introduced. Following the introduction into mathematical models of the idea that plant chemical defenses could constrain herbivore attack of plants, a more recent set of models is introduced. These are based upon the effects that plant toxins could have on the functional response of herbivores to plant biomass. This toxin-determined functional response model (the TDFRM), which has been successfully tested at least once in a long-term ecological research project, provides a potentially very powerful theoretical basis for plant chemical defense theory, especially as it applies to generalist herbivores such as browsing mammals.

The TDFRM is founded upon two observations that I made in Alaska over forty years ago. The first observation was on winter browsing mammals such as the snowshoe hare (*Lepus americanus*), the moose (*Alces alces gigas*), and ptarmigan (*Lagopus* spp.). These fed preferentially upon woody plant species, the ontogenetic stages (juvenile versus mature) of these species, and parts of the twigs of ontogenetic stages that were not rich in lipid-soluble substances that were potentially toxic. They tended to avoid eating much of the biomass of species, ontogenetic stages, and twig parts that were comparatively rich in these potential toxins. This observation suggested that the browsing mammals that I was familiar with were attempting to minimize toxin intake. The second observation was that, when a generally little-browsed species that was rich in lipid-soluble toxins, such as the Siberian green alder (*Alnus viridis* subsp. *fruitcosa*), occurred in low biomass in a forest patch, snowshoe hares browsed it to an extraordinarily high degree. I could come up with only one explanation for this observation: Even though an individual snowshoe hare could eat only a few grams of the twig biomass of green alder, if the biomass of green alder was a relatively small fraction of the forest vegetation and multiple hares each fed on the few green alder plants available to them, the combined effect of numerous hares would result in severe browsing of the few green alders. But, if the biomass of green alder was greater and the biomass of hares was constant, then, as generally observed, green alder would be lightly browsed. This observation again suggested that toxins, in this case the stilbenes pinosylvin and pinosylvin mono-methyl ether of green alder, were regulating the rate at which the herbivores, in this case snowshoe hares, were eating the biomass of their prey. If this was the case, then the rate of intake of green alder biomass by snowshoe hares could be modeled as some sort of Michaelis–Menten function in which detoxification processes were controlling the herbivore's rate of predation on its plant prey. Subsequent experiments using snowshoe hares and a toxic defense of the juvenile stage of the Alaska paper birch (*Betula neoalaskana*), the dammarane triterpene papyriferic acid, strongly supported this hypothesis.

With this information in hand, I was fortunate enough to meet with Professor Feng and to mention this possibility to her. Professor Feng immediately suggested that a good way to mathematically describe what I explained to her was to add a term to C. S. Holling's functional response predator-prey model that enabled toxicity to regulate the intake of plant biomass by a generalist herbivore. This was the beginning of the TDFRM theory developed in later chapters of this book. Subsequent to the building of the initial TDFRM, the effects of predators of herbivores such as wolves in a tritrophic system were developed, and the results of this extension now appear to accurately predict the dynamics of a woody plant-moose-wolf system in interior Alaska. Additionally, the notion of herbivore evasion of their predators has been coupled to the initial TDFRM, and this coupling could well provide a powerful tool in analyzing the "landscape of fear" hypothesis that predicts that, at the level of the landscape, toxin-determined foraging and the fear of predation interact to determine the foraging behavior of herbivores.

So, to summarize, this book begins with an excellent introduction of predator-prey theory as it applies to plant-herbivore interactions and ends with what now appears to be a powerful mathematical model of how plant toxins affect the dynamics of these interactions at levels extending from individual plant parts and individual herbivores to tritrophic interactions across entire landscapes.

John P. Bryant

Cora, Wyoming (Institute of Arctic Biology, University of Alaska Fairbanks, retired)

Preface

This book arose out of a long collaboration between the authors on attempting to use mathematical modeling to describe and understand the effects that plant defenses have on plant-herbivore dynamics. The core of the book involves a toxin-determined functional response model (the TDFRM) that was formulated with specific reference to mammalian browsers in the boreal forest confronted with plant communities in which species could have varying degrees of defense. This model and its elaborations itself spans a great range of dynamic behaviors. However, we felt it was not enough to constitute a complete book. Therefore, we have expanded the book both to include other plant-herbivore work we have been involved with and to provide an even broader context of modeling plant-herbivore interactions.

The book is divided into two halves, one a mathematical overview and the other selected applications. We begin in Chapter 1 with a very general conceptual overview of the modeling of plant growth and resource allocation, as well as of herbivore foraging, and then briefly review the resultant plant-herbivore interactions. Chapter 2 derives the basic Holling type 2 functional response and some of the general properties of predator-prey interactions with the functional response. In Chapter 3, well-known ecological models are used to illustrate five key concepts in herbivore-plant interactions. The TDFRM is described in detail in Chapter 4, including extension to spatial situations.

The applied half of the book begins with models related to a plant's dealing with herbivory, both through allocation of energy to inducible defenses and its ability to compensate for various levels of herbivory (Chapter 5). In Chapter 6, the emphasis is on herbivores' foraging strategies in response to the problems posted by low plant quality (low nutrient concentration) toxins, and predators. The use of the TDFRM to describe effects of toxicity at the food chain and ecosystem levels is covered in Chapter 7. In Chapter 8, we try to provide a broader conceptual view of how the prevalence of fire is related to the strong presence of plant toxicity in the boreal biome and how this shapes species distributions. Chapter 9 is a primer on the use of *Mathematica* in simulating the models described here. Particularly, we demonstrate the feature that allows the simultaneous visualization of model outcomes as parameter values are varying, which is especially useful for decision making in management.

This book is intended for graduate students and others who have some background in nonlinear differential equations, but we hope that the material in Chapters 2 and 3 is a relatively easy introduction that will make the rest of the book accessible to many readers. The book is not intended to be a complete textbook, as the topics by no means cover all the vast field of modeling of plant-herbivore interactions but to some extent reflect both the authors' primary experience with mammal browsers in the boreal forest. Also, we have generally avoided large, complex simulation models in favor of mathematical models of moderate complexity. But many of the key ways that nonlinear differential equations are used to describe plant-herbivore interactions are represented here.

We are indebted to our many collaborators on earlier works and publications, some of which are represented here. The TDFRM initially was developed as a collaboration between John Bryant, Zhilan Feng, and Robert Swihart, motivated by John's conjectures based on field observations and experiences in real ecological systems. This collaboration was later joined by Donald DeAngelis and Rongsong Liu. An NSF grant that supported this project (DMS-0920828) helped to establish the collaboration with a team from the University of Alaska in Fairbanks, including F. Stuart Chapin III, Tim Glaser, Knut Kielland, Mark Olson, and Jennifer Schmidt, and to provide support for students at Purdue University including Jorge Alfaro-Murillo, Matthew Barga, Muhammad Hanis B. Ahamad Tamrin, and Yiqiang Zheng. The inducible defense modeling described in Chapter 5 was the result of a collaboration of DeAngelis and a team of empiricists and modelers led by Matthijs Vos. The snowshoe hare dynamics modeling was done with Rongsong Liu, Stephen Gourley, and John Bryant. A model of plant compensation was the work of DeAngelis with Shu Ju. Some of the results for the TDFRM described in Chapters 4 and 7 involved collaborations of DeAngelis and Feng with Carlos Castillo-Chavez, Xiuli Cen, Wenzhang Huang, Ya Li, Zhipeng Qiu, and Yulin Zhao.

List of Figures

List of Tables

Part I

Basic Theory and Simple Models

Chapter 1

Introduction

1.1   Types of plant-herbivore interactions; e.g., plant and herbivore diversity, types of herbivore feeding on plants

The interaction between herbivores and plants is one of the most fundamental processes in ecology and has been the object of scientific observation and theory as early as Aristotle [352], but mathematical modeling has been applied to plant-herbivore interactions for only a few generations. The earliest mathematical models in ecology were the so-called predator-prey models of Lotka [230] and Volterra [371]. These are models described by differential equations for both populations, including terms reflecting the rate of predation of the predator on the prey, and in which the variables could be either population numbers or biomasses. Lotka–Volterrra models could describe in a qualitative way the sort of periodic oscillations that were sometimes observed in nature in terms of predator-prey interactions.

Over time many variations on the initial Lotka–Volterra equations have been proposed and analyzed to try to better understand population interactions. With little change, these predator-prey models have been adapted to describe the special class of predator-prey interactions between plants and herbivores. (The term "predator-prey" is probably not a good one for these interactions, as it conjures up an image of a predator chasing prey individuals, which does not fit herbivore-plant interactions. But we will use the term in places where general interactions are meant.)

Both theory and modeling of plant-herbivore systems have extended far beyond the simple interactions of the Lotka–Volterra type to encompass a whole range of aspects of such interactions, and these continuous-time models have been joined by discrete-time difference equation models. Even to attempt to list the manifold types of models devoted to plant-herbivore interactions would be difficult, but some sort of categorization is needed. As a first cut, Caughley and Lawton [55] divided plant-herbivore interactions into "non-interactive" and "interactive." The non-interactive type encompasses situations in which the feeding by the herbivore generally does not influence vegetation growth. Herbivory on seeds or fruits would fall into this category. Even though frugivory and seed predation can remove biomass from the plant, they do not influence the plant growth rate the way removal of foliage and roots does. Non-interactive herbivory is important ecologically, as it sustains many animal species, but it is less interesting from a mathematical standpoint than when the herbivory is interactive.

Interactive herbivory refers to situations where the rate of change of herbivore biomass depends on plant biomass and the rate of change of plant biomass depends on herbivore

Plant Actions
Allocation of carbon and nutrients to compensate for losses to herbivory
Changes in plant quality or stoichiometry
Allocation of carbon and nutrients to constitutive and induced defenses

Herbivore Actions
Patch choice, patch giving up time
Tradeoff behaviors in movement patterns in response to predators
Plant diet selection to maximize growth and balance nutrient intake
Plant selectively to minimize toxin intake
Mass attacks to counter plant anti-herbivore defenses

Effects of Plant-Herbivore Interactions
Population cycles
Herbivore outbreaks
Collapse of population (regime shift)
Maintenance of sharp vegetation or grazer ecotones
Clumping patterns of vegetation
Spacing of trees in tropical forests
Nutrient patchiness

FIGURE 1.1: Some of the responses of plants to herbivores, responses of herbivores to plants, and the results of interactions that have been modeled.

biomass; that is, there is a feedback loop. This feedback is a negative one, as the plant has a positive effect on herbivore growth rate and the herbivore has a negative effect on the plant growth rate. The theory and modeling here will be based either directly or implicitly on interactive relationships, although in some cases models will focus more on one of the interactors (plant or herbivore) than the other, with the assumption that the other interactor is influencing its actions.

No model can include all of the complexity of an interaction such as that of plants and herbivores. Modelers make deliberate simplifications of reality in their models to try to understand specific aspects of the interaction. Therefore, a variety of model types focus on parts of the whole system. An attempt to overview some of this variety is shown in Figure 1.1. Three very broad categories are shown. One model type describes the way that plants behave, especially in the way that they allocate their resources to deter herbivores or compensate for herbivory. A second is the category of foraging modeling, which includes the strategies of movement and diet selection. The third type describes the effects emerging from the interactions, such as temporal dynamics (e.g., population cycles), spatial patterns, and ecosystem functioning. These are loose categories and not all models fit conveniently into one or the other type, but we will describe these in more detail with some examples, as well as provide an overview of the rest of this book.

1.2   Plant growth and allocation

The mathematical modeling of ecological systems has generally aimed at simplicity. In this spirit, in herbivore-plant models, the dynamics of vegetation is often modeled using the simple logistic function:

$$\frac{dP}{dt} = rP\left(1 - \frac{P}{K}\right) \tag{1.1}$$

where P is the vegetation or plant biomass density (say kilograms per square meter, kg m^{-2}), r is the maximum growth rate (say kilograms per kilogram standing biomass per year, kg kg^{-1} yr^{-1}), and K is the equilibrium value that the vegetation can attain. A number of elaborations on this logistic model are discussed in Turchin [361], including dividing P into aboveground and belowground biomass. The units for vegetation are usually the density of biomass (or sometimes carbon or energy), not numbers of individuals, as terrestrial plants are seldom completely consumed by a herbivore, but merely grazed. So it makes sense to model the biomass of vegetation in a unit area.

But simple models are limited in what they can help explain, and more realism is often needed for models to be useful for many objectives. Most terrestrial woody plants are composed of the following components: Foliage (needed to capture energy and create carbon compounds), structures related to reproduction (pollen, fruits, etc.), branches (for supporting foliage), boles (both for support and to gain height advantage), roots (needed to capture nutrients). They also have mutualist organisms, which can include pollinators, seed dispersers, and myccorhizal fungi, and they produce secondary chemicals (phenolics, alkaloids, terpenes) for defense against herbivores and other purposes. The way in which the plant's energy and nutrients are allocated is called a strategy, and plants have evolved strategies to maximize their fitness. To describe vegetation in this degree of detail requires equations for the various components that are included in the model. For example, if the division of plants into foliage, fine roots, and wood (bole, branches, and the woody parts of roots) is useful for some purpose, then the equations for carbon in these compartments could be in the form

$$\frac{dC_f}{dt} = \eta_f G - \gamma_f C_f \tag{1.2a}$$

$$\frac{dC_r}{dt} = \eta_r G - \gamma_r C_r \tag{1.2b}$$

$$\frac{dC_w}{dt} = \eta_w G - \gamma_w C_w, \tag{1.2c}$$

to follow Comins and McMurtrie [72]. Here, C_f, C_r, and C_w are biomasses (or for our purposes here, carbon) of foliage, roots, and wood, respectively, the parameters γ_f, γ_r, and γ_w are senescence rates, which includes litterfall, and G is the rate of carbon fixation by photosynthesis. G is primarily a function of the amount of foliage, C_f, and environmental factors such as radiation, but depends as well on water and nutrients supplied by the roots. The parameters η_f, η_r, η_w are allocation fractions of carbon going to foliage, fine roots, and wood, where $\eta_f + \eta_r + \eta_w = 1$. This formulation of the model allows allocation to vary between components, such that optimal allocations can be found.

The plant model is made more complete by addition of equations for major nutrients, such as nitrogen, which can vary to some extent independently within certain ranges in the plant components. To include nutrients that can vary in their ratio with carbon, additional equations may be needed, again similar to those used by Comins and McMurtrie [72];

$$\frac{dN_f}{dt} = U \frac{\eta_f}{\eta_f + \rho\eta_r} - \gamma_f N_f \tag{1.3a}$$

$$\frac{dN_r}{dt} = U \frac{\rho\eta_r}{\eta_f + \rho\eta_r} - \gamma_r N_r \tag{1.3b}$$

$$\frac{dN_w}{dt} = \eta_w \nu_w G - \gamma_w C_w, \tag{1.3c}$$

where ρ is the ratio of fine root nitrogen to carbon (N:C) ratio to foliage N:C ratio. The uptake, U, is a function of fine root carbon and environmental conditions, such as nutrient availability in the soil, which may be described by equations as well. The photosynthesis rate, G, is allowed to depend on the ratio $N_f : C_f$, as well as on C_f itself in the foliage. This set of equations could be extended, if desired, to include allocation to chemical defenses,

$$\frac{dT_f}{dt} = \eta_T G - \gamma_f T_f,$$

where T_f is toxin amount in the foliage and η_T is the allocation to chemical defenses.

These equations are only a start on how the plant component of the herbivore-plant interaction might be modeled. It at least allows the modeler to explore how the plant might optimally allocate its carbon and nutrients. Models such as those of Rauscher et al. [302], Rastetter et al. [301], Comins and McMurtrie [72], Gayler and Priesack [120], and others contain greater levels of complexity.

With at least the modest level of complexity above, important aspects of herbivore-plant interactions can be studied, some of which we will consider in more detail in later chapters. One of the main questions that models that differentiate between plant components is how carbon is allocated between shoot (foliage) and root [85, 194] including allocation of nitrogen [12, 194, 355].

Other models have focused on allocation strategies that maximize carbon gain [48, 100], maximizing compensation to effects of herbivory [179], testing theories on production of secondary carbon defense compounds [120], maximizing growth of an agricultural crop [379], producing an evolutionarily stable strategy in competition with other trees [91], determining changes in allocation to chemical defenses as a function of age [29], and determining effects of differences in plant quality (nutrient to carbon ration) on plant-herbivore dynamics [229]. A large number of allocation models, including some of the above, are appraised in Le Roux et al. [212] and Franklin et al. [110].

In Chapter 5, plant allocation in relation to defense against herbivory and compensation from herbivore damage are described using the models of Gayler and Priesack [120], Comins and McMurtrie [72], and Ju and DeAngelis [179].

1.3   Modeling herbivore foraging strategies

Models of herbivores have fallen into two major types: discrete-time models that are formulated as difference equations and continuous-time models that are formulated as differential equations. The former type are typically used for describing the dynamics of univoltine insects feeding on plants; that is, insects with non-overlapping generations, such that a new generation replaces the old at regular time intervals. Differential equations are more suitable for the situation where herbivory is continuous rather than changing sharply periodically. This is certainly typical of vertebrate herbivores and can apply to insects as well in many situations.

For modeling of herbivory, a starting point is usually the formulation of a functional response of a continuous herbivore. The functional response describes the way the feeding rate of a single individual consumer responds when the density of its resource changes. Generally, the rate of feeding of the individual herbivore will increase as more edible plant biomass becomes available, but, of course, there is an upper limit on the rate at which a herbivore can feed. Therefore, a functional response must eventually "saturate"; that is, approach an asymptote as edible biomass increases toward the limit of the herbivore's ability to ingest or digest plant biomass. This asymptote is the maximum rate at which the individual herbivore can feed off the prey. A simple way of describing this saturation is hyperbolic,

$$f(P) = \frac{aP}{b + P},$$ (1.4)

where $f(P)$ is the functional response, or rate of feeding, depending on vegetation density, P, and a and b are parameters, where a is the maximum rate that vegetation can be consumed by an herbivore and b is the half-saturation constant, which corresponds to the density of vegetation at which the feeding rate reaches half its maximum.

The rate of feeding by the herbivore population (with density H) is $f(P)H$, and thus depends on the product of the herbivore biomass density and the functional response. Expression (1.4) represents only the simplest version of a functional response. The functional responses used in actual models in ecology can depend on the details of a situation, and so have taken on a number of forms, depending on the specific situation (e.g., taking into account bite size; see Cohen et al. [69]). A detailed derivation of some common functional responses is given in Chapter 2. Comprehensive treatments of the modeling of foraging exist, such as those for vertebrate herbivores by Owen-Smith [280], Fryxell and Lundberg [116], and Crawley [74], as well as many that focus on the functional response, e.g., Lundberg [233].

The various aspects of foraging in a landscape context were outlined conceptually by Senft et al. [327]. We briefly note how modeling has been adapted to the strategies of herbivore foragers faced with landscapes with arrays of opportunities and constraints. Landscapes are usually heterogeneous, or patchy, with respect to resources like edible vegetation, so the strategy of movement between patches (the choices of patches to visit and how long to forage in them) has been of interest to modelers such as Charnov [63], Searle et al. [325], and de Knegt et al. [79]. Spalinger and Hobbs [338], Gross et al. [133], and Hobbs et al. [155] show that the spatial arrangement of patches influences the functional response, explaining some of the empirically observed variation, although the nature of the response has the same hyperbolic form as in expression (1.4). The distribution on the landscape of other needs, such as water and shade [156] influence movement patterns, as does the distribution of risk of predation [35, 225] and parasitism [163]. The spatial mixture of plants of different species may affect the amount of herbivory on focal plants [139].

Different plant species, as well as different ages and different parts of individual plants, have different nutritional values, so diet selection has been a major focus of modeling (e.g., [25, 106, 147, 161, 279, 332]). The heterogeneous distribution of vegetation quality leads to movement behaviors, often based on memory of reliable foraging areas [104], and to patterns of aggregation of herbivores [115]. Plant chemical defenses have evoked responses in herbivores such as mass attacks, which have been modeled [228]. Selective feeding, even down to the selection of different parts of twigs on the same woody plant, has been described in models considered in [81, 227].

We have only touched on the vast literature on the subject of herbivore foraging. Models of Owen-Smith and Novellie [279] and Belovsky [25] will be considered in more detail in Chapter 6.

1.4   The consequences of herbivore-plant interactions

The interactions between plants and herbivores create some of the fundamental ecological patterns in time and space. These have been elucidated by both mathematical and simulation models. A few types of patterns are described here, some of which are described in more detail in later chapters.

A temporal pattern that has been of great interest to theoretical ecologists is the periodic oscillation, or cycling, of populations. These can be exogenously driven, such as when they respond to annual cycles of environmental conditions. But they can also be endogenously driven, when mechanisms within the ecological system in which the population exists creates its own ability to cycle without continual external driving. Herbivore-plant interactions contain mechanisms that can lead to cycling under certain conditions. Fundamentally, the plant component is autocatalytic; that is, when well below its carrying capacity, it grows at a rate proportional to its biomass. Second, the herbivore population grows by feeding on the plant population, and thus acts as an inhibitor, or a negative feedback, on plant population growth. Third, there should be some time lag, or delay in the negative feedback's having an effect. In some models, there is a delay in the reduction in plant quality following herbivory [13, 92]. Heavy herbivory on the plant population can result in a shift toward lower nutritional value in the plant population by removing the more nutritious plants preferentially. This creates a delayed negative effect on the herbivore population, which leads to a rebound of the plant population, followed by a rapid increase in herbivores and a continuation of the cycles. Also, heavy herbivory in one year on vegetation can stimulate production of a high concentration of plant secondary chemicals the following year, which acts as a delayed negative feedback on the herbivores (e.g., [81, 93, 227]). The presence of a constitutive, or permanent, chemical defense can also cause delayed feedbacks on growth and survival of herbivores. A "toxin-determined functional response" [97] for this type of defense is described in Chapter 4, while the dynamics resulting from induced defenses are modeled in Chapter 5. Other hypotheses for cycling, based on fluctuations in biomass alone, rather than changes in plant quality, are supported by models for some situations [1]. The role of changes in plant quality on cycling is considered in Chapter 3 in a model of plant quality in a plant-herbivore interaction by Loladze et al. [229].

Outbreaks of herbivores are phenomena that may occur as cyclic patterns, as above, but can also be irruptive, triggered by factors that are irregular in time. Some of the irruptive outbreaks may be explained by a so-called regime shift, in which slowly changing environmental conditions lead to a sudden release of the herbivore population from the state of being under control by natural enemies to the state in which it grows in an uncontrolled manner. The release phase is crucial, allowing it to have catastrophic effects on the vegetation. This is clearly something that foresters and ecologists alike want to understand. The model of Ludwig et al. [232] for outbreaks of the spruce budworm is of this type and is discussed in detail in Chapter 3. Kang et al. [182] described a model showing an outbreak similar to that of a gypsy moth population interacting with host plants. An insect outbreak model in a spatial setting was studied with a simulation model by Myers [262], who modeled larval insects in an environment of many individual plants. Each individual plant could, on average, support 10 larval insects to adulthood. Insects surviving a plant were either dispersers, in which case they laid eggs on any of the other plants, or non-dispersers, in which case their offspring remained on the same host plant. This tendency to remain on the plant could lead to local overexploitation of the plants resulting in the death of the larvae on the plant; that is, local "flame-out," paradoxically leading to greater stability over the system as a larger spatial whole.

The concept of the regime shift mentioned above applies far more broadly in plant-herbivore systems than just herbivore outbreaks. A regime shift is a major change in an ecological community that might be stimulated by an external disturbance or a gradual change in environmental conditions that take the system past a threshold beyond which a change inevitably occurs, but it is internal feedbacks that drive the system to its new state. These states are called alternative stable equilibrium states, because either state may occur at a particular location, depending on initial conditions. Both of these alternative communities (they can also be termed alternative ecosystems) are relatively stable; that is, once they occupy a site, they don't easily change, because each tends to maintain environmental conditions favorable to itself. This is termed bistability. However, a big enough disturbance, such as a fire, or a prolonged environmental change, like a gradual increase (or decrease) in temperature over many years, could cause the system to switch to the alternative stable state ecosystem. What makes the changes between these states "regime shifts" is that once conditions have passed a certain point, positive feedbacks propel the system toward a new state, and once the system reaches the new, alternative state, the feedbacks tend to keep it there.

The broad applications of the concept of regime shifts, or critical transitions, are described in [322]. A grazing system ecologist, Immanuel Noy-Meir [268], was perhaps the first to demonstrate the possible occurrences on such regime shifts in a consumer-plant (grazer-grass) model, which allows one to understand the sharp sudden tipping that has been observed in such systems. Collapse of the vegetation was one phenomenon that could be interpreted as a critical transition. This model is described in Chapter 3.

Herbivore-plant interactions are also factors in the formation of spatial patterns. Two general ways in which these spatial patterns develop are described in models in later chapters. The first model is the application of the toxin-determined functional response model (TDFRM) to the forest succession on the landscape [75] in Chapter 7. A spatially explicit simulation model is also applied to the forest succession influenced by herbivore browsing and predator control in Chapter 7. Disturbances, in this case fire, continually create new open areas on which secondary succession occurs, such that there is a heterogeneous pattern of patches in various stages of succession across the landscape. The TDFRM shows that selective herbivores can speed up the process of succession from less-defended deciduous vegetation to more protected coniferous vegetation. The TDFRM can also be used to explain how the invasion by a non-native plant species can be facilitated by herbivory. When the invading plant is more toxic than the native vegetation and the herbivore feeds selectively to avoid high levels of toxicant intake, the TDFRM based on these assumptions can have a traveling wave solution of invasive plant movement, as described in Chapter 8.

Another aspect of spatial patterns is the formation of boundaries, often sharp, between vegetation types, or even whole ecosystems, as the result of herbivore-plant interactions. We will not consider models of this type in the book but, because this phenomenon is of growing interest, we mention a few literature models here. The alternative stable states and bistability described above apply to the formation of sharp boundaries, or ecotones, between ecosystem types along a spatial gradient, say of a climatic variable, even though the spatial change of the variable is very gradual. The reason for the woodland frequently sharp ecotone is that each alternative state, once established, is self-reinforcing, and tends to prevent invasion by the other state. The savanna-tropical woodland ecotone is a case in point. The high density of grass in a savanna ecosystem favors fire, which suppresses forest tree seedlings, while high density of forest trees shades out grass and reduces the probability of fires. This creates conditions for bistability, which can lead to creation of sharp ecotones. Herbivory may also play a role in this process. In particular, elephants suppress tree establishment and growth in savannas, reinforcing the boundary between more open savanna and woodland. The situation may be even more complex. Modeling by van de Koppel and Prins [364] shows that competition of elephants and another large herbivore, giraffes, with smaller herbivores such as impala or buffalos, may increase the need of the former to exploit woody vegetation, affecting the transition from savanna to woodland.

Not only is vegetation often characterized by sharp changes along an environmental gradient, but the grazer communities themselves also can undergo sharp changes of alternative stable states, as shown in models. There are tradeoffs in which some grazers are better competitors for biomass (carbon) and others at extracting essential nutrients. Along a soil fertility gradient in which the nutrient concentration in plants changes, competing grazer species that are better adapted to exploiting nutrients will replace the other type along the direction of increasing soil fertility. Modeling shows that at intermediate soil fertility, bistability may occur, with one or the other alternative state occurring. One state is dominated by the community of grazers efficient at obtaining biomass (carbon-efficient grazers), while in the other state the nutrient-efficient grazers dominate [138]. This situation of bistability can occur if the nutrient-efficient grazer is limited by nutrient and the carbon-limited grazer is limited by carbon. In that case, either competitor could reduce the resource it consumes more efficiently to the level where the other species could not survive.

Fertility gradients can also have some paradoxical effects. Increasing soil fertility in a wetland ecosystem may lead to light becoming the limiting factor, favoring larger less palatable plant species, which reduces the carrying capacity of the system for the herbivores [366]. Herbivore grazing can affect the age and species composition of grasslands. Low levels of grazer densities can maintain the vegetation in a more juvenile and digestible state, which increases the rate of intake by the grazers [115]. Further, if the grazers select preferentially the more digestible vegetation, this positive effect on the grassland is reinforced [164].

Schneider and Kéfi [323] showed that fine-scale heterogeneity of vegetation formed by the interaction of grazers with plants, whose grazing resistance increases with clump size, determines the grazing intensity at which a regime shift can occur from vegetation to bare ground. Movement rules of individual foragers also affect the way that community structure changes under habitat loss [47]. In a model of sheep grazing of a semiarid grass steppe, Paruelo and others [283] found that the non-uniform grazing that reduces tussock density through creating a pattern of highly defoliated and non-defoliated areas, has a larger effect on aboveground annual productivity than does an overall decrease in individual tussock production. Positive feedback between grazing and plant palatability is shown through modeling to be a mechanism that can create patterns of tall plant stands (tussocks) and short grass (lawns) in savanna ecosystems [259]. The depletion of seeds by seed predators has been demonstrated by models to affect the distribution of trees in tropical forests and may help explain the relatively even spacing of trees [264]. The pattern of selective browsing by moose was shown by modeling to lead to a spatially heterogeneous pattern of browsed and unbrowsed vegetation. This leads to further ecosystem effects, as the unbrowsed vegetation has slowly decomposing litter, which leads to less soil nitrogen in those sites [286].

The above is a very brief survey to provide a general overview of some ways that models have been used to explore the interactions of plants and herbivores. We can consider only a few models in detail, but we hope these introduce some of the important ways ecologists are modeling this important interaction. We are also including a final chapter (Chapter 9), which provides examples of a few of the models described as *Mathematica* Notebooks.

Chapter 2

Predator-Prey Interactions

2.1   Beginnings of predator-prey modeling

The introduction of mathematical modeling into ecology is close to 100 years old. The first attempts at modeling the interactions between populations of predators and prey were made in the 1920s, independently in America and Italy by Alfred J. Lotka and Vito Volterra [230, 370]. Lotka expressed the interactions between predators and prey verbally:

Lotka's verbal expression for predator-prey interactions

(Change in the number of prey per unit time) = (Natural increase of prey per unit time) − (Destruction of prey by predators per unit time);

(Change in the number of predators per unit time) = (Increase in predators per unit time as a result of ingestion of prey) − (Deaths of predators per unit of time).

When written in mathematical form as rates of change in time, this verbal description becomes

$$\frac{dP}{dt} = rP - k_1 PH \tag{2.1a}$$

$$\frac{dH}{dt} = k_2 PH - mH. \tag{2.1b}$$

These coupled equations have become known as the Lotka–Volterra (LV) predator-prey equations, where P and H are densities (e.g., number of individuals per unit area) of the prey and predators, respectively, r is an intrinsic growth rate of the prey, m the mortality rate of the predators, k_1 a loss rate coefficient for the prey due to predation, and k_2 a rate coefficient of increase of the predator population due to the interaction. The dimensions of r and m are time^{-1}, while those of k_1 and k_2 are time^{-1}(number of predators)$^{-1}$ and time^{-1}(number of prey)$^{-1}$.

It has been noted [142, 303] that Lotka and Volterra derived these equations in different ways. Volterra [370] wrote down equations similar to (2.1) directly, arguing that interactions between individuals should follow a simple mass action law, being proportional to the product of the densities. Lotka doubted that specific equations for species interactions could easily be found, so began with a set of general equations for n species, for population sizes, X_i,

$$\frac{dX_i}{dt} = F_i(X_1, X_2, \cdots, X_n), \quad i = 1, 2, \cdots, n, \tag{2.2}$$

([230], p. 57), which he expanded as a Taylor series, terminating it with second-order terms and keeping only the $X_i X_j$ terms and ignoring intraspecific competition terms X_i^2. Equations (2.1a,b) are the result when this general set of equations of interacting populations is reduced to one predator and one prey. From Lotka's perspective, the equations (2.1a,b) are valid only locally around the equilibria $(0, 0)$ and $(m/k_2, r/k_1)$. Therefore, these equations are only an approximation of what holds globally over the whole range of values of X_1 and X_2.

Nicholson [267] recognized that lack of intraspecific competition in equation (2.1a) was unrealistic, as natural enemies do not always control their prey, and that, in their absence, the prey described by (2.1a) would tend toward infinity. So competition among the prey for resources must ultimately limit the prey. Intraspecific competition of prey would also limit the severity of the cycling behavior predicted by the LV equations, which are not easily found in laboratory experiments on predators and prey ([336], citing [118]). In fact, a simple equation for intraspecific competition in a population had been proposed by Pearl and Reed [289], the logistic equation

$$\Big(\text{Rate of change in the number of prey per unit time}\Big) =$$

$$\Big(\text{Maximum potential increase}\Big) \times \Big(\text{Degree of realization of potential increase}\Big)$$

or, mathematically, as

$$\frac{dX}{dt} = rX\left(1 - \frac{X}{K}\right). \tag{2.3}$$

Nicholson [267] also realized that the terms for the interaction between the species were not realistic and critiqued the earlier work as follows: "It is of interest to note that one of the fundamental hypotheses of Volterra's work is that the number of offspring produced by predators is proportional to the number of encounters with prey." Consequently, though he speaks of "eaten species" (specie mangiata), he actually deals with parasites possessing an unlimited egg supply. Predators (using the term in its generally accepted sense) do not eat prey in proportion to the number found, for hunger and satiety commonly determine whether the prey met are eaten or not. Similarly, the number of offspring produced, though it may be influenced by, is clearly not proportional to the number of prey eaten by the predator (page 148). Therefore, the simple factor, $k_1 X_1$, multiplying the prey density, could not be right, and a more realistic function, what Solomon ([336], page 16) called the functional response, had to be found. The interaction could depend on a number of factors, as recognized by Leopold ([216], quoted by Holling [158]): (1) density of the prey population, (2) density of the predator population, (3) characteristics of the prey (reaction to predators, stimulus detected by predator, etc.), (4) density and quality of alternative foods available for the predator, and (5) characteristics of the predator (food preferences, efficiency of attack, etc.).

Finding a way to sort those factors out was difficult. As Nicholson [267] pointed out, "So far it has not been possible to formulate the problem of the interaction of true predators and their prey in a completely satisfactory way for mathematical treatment" (page 148). Holling took up this search. His aim was to find a comprehensive functional response that could apply in a wide range of situations, incorporate all of the factors identified by Leopold, and yet be relatively simple. Deriving a relationship describing the rate of predation that clearly showed the effects of prey and predator densities required a simple enough situation. Holling studied predation on pine sawflies in the cocoon stage in a pine stand. Both the total number of prey and the ones that had been opened by predators (masked shrew, short tail shrew, and deer mouse) could be counted and estimates of the numbers of the predators made. From observations in the field and additional laboratory experiments, the components of predation could be measured. Other factors, such as alternative food, prey variation, predator efficiency, etc., could be ignored and the whole interaction reduced to two variables, prey and predator densities. From estimates of the numbers of predators, prey, and destroyed prey, the daily number of prey consumed per predator at different cocoon densities could be calculated. The results were that the number of cocoons eaten by individuals each day increased with cocoon density, but reached an asymptote. This can be described by the Holling type 2 functional response (see below). Holling called attention to other types of response. For example, Gause [118] and Burnett [49] found that the number of parasites (y) as a function of hosts (x) parasitized flattened out as described by $y = a(1 - e^{kx})$ similar to a Monod or Holling type 2 response. Solomon [336] coined the term functional response "As host density rises, each enemy will attack more host individuals, or it will attack a fixed number more rapidly" (page 16).

2.2 Derivation of the Holling type 2 and type 3 functional responses

The functional response found empirically by Holling can be derived mathematically by making some simple assumptions. Imagine that a predator is searching in a region in which there are prey scattered randomly about, with a spatial density, P (prey number), as shown in Figure 2.1. The question is, how many prey will the predator capture and eat per unit time, if it spends all of its time either searching for prey or eating them? We can answer this question if we can quantify the searching rate, visual range of the predators, detectability of prey, probability of the predator capturing prey items that it spots, and the time it takes the predator to consume an individual prey item.

If the linearly directed movement rate or velocity of the predator is s and it has sensing range of v to either side, the area that the predator searches during a time period T_s is

Area searched per unit time $= 2vsT_s$.

Next assume that the "prey detectability," or the probability that the predator can detect a prey item that lies within the search area, is k. Then, given that the prey density is P prey per unit area, the number, P_a, of prey detected during time T_s is

$$P_a = (\text{detectability}) \times (\text{area searched, m}^2) \times (\text{prey per unit area, m}^{-2})$$

or

$$P_a = k \times (2vsT_s) \times P = (2vsk)T_sP \equiv fT_sP,$$

where the dimensions of f are square meters per time (for search on a two-dimensional plane), such as m^2d^{-1}. But we want to calculate the number of prey that are actually captured, P_c, during that time, so we must multiply P_a by the capture probability, c;

$$P_c = cP_a = cfT_sP.$$

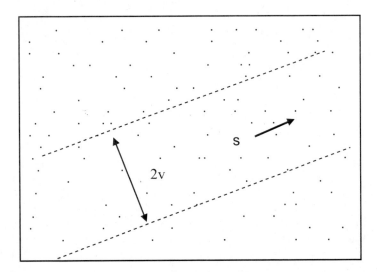

FIGURE 2.1: Predator moving at some mean "searching" speed, s, and has at least the potential to spot prey that are within a distance v on either side.

We have assumed so far that the predator spends time T_s searching. But we must also take into consideration that the predator spends time both in chasing and in consuming (which may include part or all of the process of digesting) the prey, both of which take time away from searching. Therefore, if T_c is the time spent chasing each prey individual that is detected and T_m is the time spent consuming each prey that is captured during foraging, these add to the predator's time budget.

Assume that the consumer has some amount of time available, T_{total}, each day (or other time unit) to spend in searching, chasing, and consuming prey. Then it must be true that

$$T_{total} = \text{(time spent searching)}$$

$$+\text{(time spent chasing prey} \times \text{number of prey detected)}$$

$$+\text{(time spent consuming prey} \times \text{number of prey captured)}$$

or

$$T_{total} = T_s + T_c P_a + T_m P_c,$$

which can be rewritten as

$$T_{total} = T_s + T_c P_a + T_m c P_a = T_s + (T_c + T_m c)P_a.$$

Define T_h to be the total "handling time," which included chasing time plus consumption time; that is, $T_h \equiv T_c + cT_m$. Henceforth, we will use the conventional symbol h for the prey handling time, instead of T_h. We will also consider the capture probability, c, to be 1 and chasing time $T_c = 0$, as we will specialize predator-prey interactions to herbivore-plant interactions. Then the above equation can be written in more abbreviated form as

$$T_{total} = T_s + hP_a.$$

Solving for T_s, we obtain

$$T_s = T_{total} - hP_a,$$

and using

$$P_c = P_a = fT_s P$$

in the preceding expression and rearranging, we obtain

$$T_s = \frac{T_{total}}{1 + fhP}$$

so that the number of captures per unit time is

$$P_c = fPT_s = \frac{fPT_{total}}{1 + fhP}.$$

Consider what this means. P_c represents the number of prey captured during a time period T_{total}. To obtain the rate of prey captures, one must divide by T_{total}, the total time available for feeding activities during a period of time (such as a day). So, dividing P_c by T_{total}, we obtain

$$\text{Rate of prey captures per predator } (d^{-1}) = \frac{P_c}{T_{total}} = \frac{fP}{1 + fhP}. \tag{2.4}$$

This is the Holling type 2 functional response (HT2FR) Note that it implies that, as the prey density, P, increases, the rate of prey captures also increases. However, this rate approaches the asymptote of $1/h$ as P increases to very large values.

This can be changed to the more familiar form of the Michaelis–Menten equation by dividing the numerator and denominator by fh. We then obtain

$$\text{Rate of prey captures per predator } (d^{-1}) = \frac{(1/h)P}{\frac{1}{fh} + P}.$$

Note that the above equation now looks like the Michaelis–Menten equation, except for the following changes:

Michaelis–Menten	Holling type 2
Resource density, S	Prey density, P
Maximum uptake rate, P_{max}	Maximum prey consumption rate, $\frac{1}{h}$
Half-saturation const., K_s	Half-saturation const., $\frac{1}{fh}$

The reason for the asymptotic behavior is that eventually, no matter how many prey are available, the predator is going to be limited by the time it takes to chase and consume the prey, so the absolute upper limit on the number of prey that it can take during a time period, such as a day, is

$$\text{Maximum number of prey consumed per unit time } = \frac{1}{h}$$

The meaning of h should be discussed, because it may seem a bit vague as to what it actually means. It was defined above as the "time required to consume a prey individual." However, when talking about long-term consumption rates, it should be more precisely the "mean time required to handle the consumption and digestion of an individual prey."

The Holling type 3 functional response can be derived by altering an assumption in the above derivation. We assumed that the "prey detectability," or the probability that the predator can detect a prey item that lies within the search area, is a constant, k. However, behavioral studies [358] have shown that the predator's ability to detect prey may increase with prey density, v, as the predator more easily forms a "search image" of the prey. Assuming detectability increases linearly with P, then $v = v_a P$, where v_a is a constant,

$$P_a = k \times (2 v_a P_s T_s) \times P = (2 v_a s k) T_s P^2 \equiv f T_s P^2$$

and, using the above approach,

$$\textit{Rate of prey captures per predator } (d^{-1}) = \frac{fP}{1 + fhP^2}. \tag{2.5}$$

The Holling type 2 functional response has also been extended to include the effect of predator interference [24, 82]. This extension can be derived by assuming that predators searching for prey encounter each other at some rate proportional to predator density, $k_w H$ (where H is predator density) and each encounter occupies some time T_w. Then the total time can be written as

$$T_{total} = T_s + fhT_s P + k_w T_w H T_s.$$

Then

$$\textit{Rate of prey captures } (d^{-1}) = \frac{P_c}{T_{total}} = \frac{fP}{1 + fhP + k_w T_w H}.$$

2.3   Incorporating the predator functional response into one-predator and one-prey equations

Recall that Nicholson [267] recognized that lack of intraspecific competition in equations (2.1a,b) was unrealistic, as competition among the prey for resources must occur in the limit of a large prey population. First, we want to describe the resource-limited growth of the prey population. The simplest description of such growth has traditionally been the logistic equation. Assume that P represents the number density of the prey and also assume that either nutrients or light is limiting, such that the biomass density of prey (g m^{-2}) has an upper limit of K. Assume also that the maximum per capita growth rate of the prey, when the prey population is small, is r. Then the logistic equation describing growth of biomass has the form (2.3), i.e.,

$$\frac{dP}{dt} = r\left(1 - \frac{P}{K}\right)P.$$

[There have been objections to the use of this form of the logistic equation, which is also called the Pearl–Verhulst equation, as r and K are not independent in this form, and difficulties occur when additional mortality is imposed (see [15, 239]). The original form introduced by Verhulst,

$$\frac{dP}{dt} = rP - aP^2$$

is suggested by the critics to be superior. We will stick with (2.3) and avoid situations that cause difficulty.]

Notice that the equilibrium density of prey in (2.3) is $P^* = K$. If we increase K, by increasing the amount of light or nutrients (whichever was limiting), then P^* increases. This is an example of "bottom-up" control, in which the size of a population or trophic level is controlled by the level of its resources.

Now consider the case in which a predator feeds on the prey. Recall that the Holling type 2 functional response for the rate of prey captures by a single predator is given by

(2.4). If the population density of predators is H, then the rate of prey captures by the predator population is

$$\text{Rate of prey captures } (d^{-1}) = \frac{fPH}{1 + fhP}. \tag{2.6}$$

It may be convenient to express the interaction of prey and predator in terms of biomasses, rather than numbers, as above. To obtain equations in terms of biomass, we define

b_p = mean biomass of a prey individual

b_h = mean biomass of a predator individual

Therefore, if P_B is the biomass density of the prey and H_B is the biomass density of the predator, we have

$$P = \frac{P_B}{b_p}, \quad H = \frac{H_B}{b_h}.$$

Then we can rewrite (2.6) as

$$\text{Rate of removal of prey biomass } (d^{-1}) = \frac{fP_BH_B}{1 + fh(P_B/b_p)} \cdot \frac{1}{b_pb_h}.$$

We can now write the equation for the prey (numbers in this case) by subtracting the rate of removal by predators, Holling type 2 in this case, or (2.6), from the prey logistic equation (2.3);

$$\frac{dP}{dt} = r\left(1 - \frac{P}{K}\right)P - \frac{fPH}{1 + fhP}. \tag{2.7a}$$

We need to complete the model by writing down an equation for the rate of change of H,

$$\frac{dH}{dt} = \frac{\gamma fPH}{1 + fhP} - mH, \tag{2.7b}$$

where γ represents conversion of prey to predator number and $-mH$ represents the loss rate of predator biomass due to death and respiration of predator. Equations (2.7a) and (2.7b) constitute the description of our predator-prey system. The corresponding equations for number densities are

$$\frac{dP_B}{dt} = r\left(1 - \frac{P_B}{Kb_p}\right)P_B - \frac{fP_BH_B}{1 + fh(P_B/b_p)} \cdot \frac{1}{b_h} \tag{2.8a}$$

$$\frac{dH_B}{dt} = \frac{\gamma_B fP_BH_B}{1 + fh(P_B/b_p)} \cdot \frac{1}{b_h} - mH_B, \tag{2.8b}$$

where γ_B represents conversion of prey biomass to predator biomass.

Both of these pairs of equations (2.7) and (2.8), and variations on them, have been the object of much study in theoretical ecology. Our goals here are limited. We just want to see what the equations imply about the equilibrium sizes of the two trophic levels and then study the stability of the system. In what follows we will use P and H to refer to number or biomass, depending on the model.

2.3.1   Equilibria of the one-prey one-predator system

The equilibria for the model equations (2.7a) and (2.7b) can be found in the (P, H)-plane, by plotting the zero isoclines, which are found by setting the right-hand sides of (2.7a) and (2.7b) to zero and plotting H against P. The zero isoclines are:

$$P = \frac{m}{f(\gamma - mh)}, \tag{2.9a}$$

$$H = \frac{r}{f}\left(1 - \frac{P}{K}\right)(1 + fhP). \tag{2.9b}$$

The predator zero isocline (2.9a) is a vertical line in the plane, while the prey zero isocline (2.9b) has a hump shape with the peak occurring at the point $P_{peak} = K/2$ (Figure 2.2). The interior equilibrium point is formed by the intersection of the two zero isoclines, while boundary equilibria occur as $(0, 0)$ and $(K, 0)$ as well. The solutions for the interior equilibrium is found from (2.9a) and (2.9b):

$$P^* = \frac{m}{f(\gamma - mh)}, \tag{2.10a}$$

$$H^* = \frac{r}{f}\left(1 - \frac{P^*}{K}\right)(1 + fhP^*). \tag{2.10b}$$

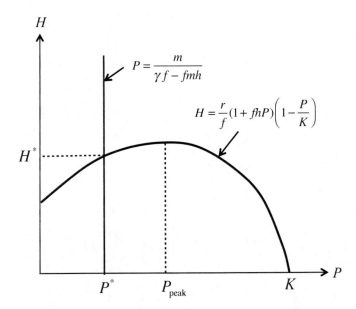

FIGURE 2.2: Diagram of predator-prey state plane.

Note that the equation for P^* (2.10a) has no dependence on K. It depends on such predator parameters as h, m, and f. What this means is that the prey population is controlled by the predator population, not the level of resources, K. Let us now look at the expression for the predator. We see that it *does* depend on K. Notice first, that if K is sufficiently small, then it may be true that $1 - P^*/K < 0$, which would mean that $H^* < 0$. What that means is that the predator population cannot exist unless K is large enough. That is, it is necessary for

$$K > \frac{m}{f(\gamma - mh)}$$

for the predator population to exist. Of course, if the predator population does not exist, then the prey population is described by equation (2.3) and $P^* = K$. There is a threshold value in the limiting resources for the predator. Below that threshold, the predator cannot exist. In this region of values of K, the equilibrium density of the prey increases linearly with K;

$$P^* = K.$$

As K is increased from very small values, it will eventually reach the point where

$$K = \frac{m}{f(\gamma - mh)}.$$

As soon as K passes this threshold, the predator can invade. Then, as K is increased further, the prey density, P^*, stays the same; $P^* = \frac{m}{f(\gamma - mh)}$. It is held by the "top-down" effect of the predator. The predator, on the other hand, continues to increase as K is increased further, though asymptotically. So we see that the dependence of the prey biomass density changes from dependence on its resources (bottom-up) to dependence on the predator (top-down).

2.4 Functional response for predation on two or more prey species

Recall that the area searched by the predator during a time period T_s is

Area searched per unit time $= 2vsT_s$

The prey detectability, k, will differ for different prey types. Simplifying to two prey types, we have k_1 and k_2. Also, we introduce the assumption that the predator can make a choice of whether to attack a particular prey encountered; it may prefer one prey type to the other. So we introduce the parameters, σ_1 and σ_2 to represent the probabilities that the predator will attack a prey of a given type; that is, selectivities. Then, given that the prey density of prey type 1 is P_1 prey per unit area, the number, P_a, of prey detected during time T_s is

$$P_{1a} = (selectivity) \times (detectability) \times (area\ searched,\ m^2) \times (prey\ per\ unit\ area,\ m^{-2})$$

or

$$P_{1a} = \sigma_1 \times k_1 \times (2vsT_s) \times P_1 = \sigma_1(2vsk_1)T_sP \equiv \sigma_1 f_1 T_s P_1,$$

and

$$P_{2a} = \sigma_2 f_2 T_s P_2,$$

$$T_{total} = (time\ spent\ searching) + (time\ spent\ consuming\ prey)$$
$$\times (number\ of\ prey\ 1\ captured + number\ of\ prey\ 2\ captured)$$

or

$$T_{total} = T_s + h_1 P_{1c} + h_2 P_{2c}.$$

Using $P_{1a} = \sigma_1 f_1 T_s P_1$ and $P_{2a} = \sigma_2 f_2 T_s P_2$ and solving for T_s, we obtain

$$T_s = \frac{T_{total}}{1 + \sigma_1 f_1 h_1 P_1 + \sigma_2 f_2 h_2 P_2}$$

and

$$P_c = \frac{\sigma_1 f_1 P_1 + \sigma_2 f_2 P_2}{1 + \sigma_1 f_1 h_1 P_1 + \sigma_2 f_2 h_2 P_2}.$$

When this is inserted into a model of a predator and two prey species, the equations are

$$\frac{dP_1}{dt} = r_1\left(1 - \frac{P_1}{K_1}\right)P_1 - \frac{\sigma_1 f_1 P_1 H}{1 + \sigma_1 f_1 h_1 P_1 + \sigma_2 f_2 h_2 P_2} \tag{2.11a}$$

$$\frac{dP_2}{dt} = r_2\left(1 - \frac{P_2}{K_2}\right)P_2 - \frac{\sigma_2 f_2 P_2 H}{1 + \sigma_1 f_1 h_1 P_1 + \sigma_2 f_2 h_2 P_2} \tag{2.11b}$$

$$\frac{dH}{dt} = \frac{\gamma(\sigma_1 f_1 P_1 + \sigma_2 f_2 P_2)H}{1 + \sigma_1 f_1 h_1 P_1 + \sigma_2 f_2 h_2 P_2} - mH, \tag{2.11c}$$

where P_1 and P_2 are the two prey species.

2.5   Stability of model equilibria

An important question is whether an equilibrium point of the predator-prey system is stable or unstable. It can be easily shown that the equilibria $(0,0)$ and $(K,0)$ of (2.7a,b) are both always unstable, so we will consider only the equilibrium (P^*, H^*).

If an equilibrium point is stable, then any perturbation away from the equilibrium point will lead to a return, whereas, if it is unstable, the trajectory will spiral away toward a stable limit cycle, in which both populations exhibit population cycles. To evaluate local stability, one can assume small perturbations around the equilibrium point by setting

$$P = P^* + x, \quad H = H^* + y,$$

expanding the right-hand sides of (2.7a) and (2.7b) in x and y, keeping only first-order terms in x and y. We obtain the equations

$$\frac{dx}{dt} = \left[r - \frac{2rP^*}{K} - \frac{fH^*}{(1 + fhP^*)^2}\right]x - \frac{fP^*}{1 + fhP^*}y \equiv c_{11}x + c_{12}y$$

$$\frac{dy}{dt} = \frac{\gamma f P^*}{1 + fhP^*}x = c_{21}x.$$

These equations are linear, and the eigenvalues can be found from the equation

$$\lambda^2 - c_{11}\lambda - c_{12}c_{21} = 0$$

where

$$c_{11} = r - \frac{2rP^*}{K} - \frac{fH^*}{(1 + fhP^*)^2}, \quad c_{12} = -\frac{fP^*}{1 + fhP^*}, \quad c_{21} = \frac{\gamma fH^*}{1 + fhP^*}, \quad c_{22} = 0$$

or

$$\lambda_{1,2} = \frac{1}{2}c_{11} \pm \frac{1}{2}\left(c_{11}^2 + 4c_{12}c_{21}\right)^{1/2}.$$

Note that, because $c_{12} < 0$, the sum of the terms in the argument of the square root is less than c_{11}^2, so λ_1 or λ_2 can only be positive (instability), if

$$c_{11} = r - \frac{2rP^*}{K} - \frac{fH^*}{(1 + fhP^*)^2} > 0$$

or, using

$$H^* = \frac{r}{f}\left(1 - \frac{P^*}{K}\right)(1 + fhP^*),$$

if

$$r - \frac{2rP^*}{K} - \frac{r(K - P^*)}{K(1 + fhP^*)} > 0.$$

From this, it can be shown that it is necessary for instability that

$$P^* < \frac{K}{2} - \frac{1}{2fh}. \tag{2.12}$$

It can further be shown that this expression is the point where the peak of the humped curve for the prey isocline (2.9b) occurs (see Figure 2.2). It can be obtained by differentiating (2.9b) with respect to P, and setting the result to zero to obtain

$$P_{peak} = \frac{K}{2} - \frac{1}{2fh}.$$

In this chapter we have derived some basic relationships of predator-prey models. In Chapter 3 we specialize to herbivore-plant models and derive a number of results.

Chapter 3

Overview of Some Results of Plant-Herbivore Models

3.1   Introduction

The purpose of this chapter is to briefly introduce some basic theory of plant-herbivore interactions through several models from the plant-herbivore literature.

When ecologists began applying models to plant-herbivore interactions, they started with models similar to the general predator-prey models (Lotka–Volterra) described in Chapter 2, the variables now being plant and herbivore biomasses. But these were specialized in some ways. Caughley [54] first categorized plant-herbivore interactions into "non-interactive" and "interactive."

3.1.1   Non-interactive

In this category of models, plant or vegetation growth is uninfluenced by consumers. That is, the herbivore-vegetation "interactions" are special in that they do not remove vital biomass from the plants. For example, many granivorous, or seed-eating rodents do not directly harvest seeds from plants, but depend on those that fall from the trees or shrubs [296, 310]. The herbivore does not subtract biomass from the plant and therefore does not influence the growth rate or size of the vegetation. The herbivore may not even care what the sizes of the plants are, if it depends only on the rate of primary production. The equations below can describe this situation;

$$\frac{dP}{dt} = rP\left(1 - \frac{P}{K}\right) - fP \tag{3.1a}$$

$$\frac{dH}{dt} = \gamma f P - mH \tag{3.1b}$$

where P is plant biomass density, and H herbivore biomass density. The plant biomass is described here by a logistic function, which was described in Chapter 2. The density dependence in the logistic function represents that because resources, such as light, water, nutrients, and CO_2 are limited, the growth rate of the plant biomass slows relative to losses such as respiration as biomass density increases. The additional term for loss rate, fP, accounts for losses of biomass through dropping of leaves and seeds which are available for ground granivores. It is assumed that usually $r > f$. Consumption is purely dependent on the amount of production that falls to the ground. There is no feedback effect of the herbivore consumers on the plant. The equilibrium values of vegetation and herbivores are

$$P^* = \frac{K}{r}(r - f), \quad H^* = \frac{\gamma f P^*}{m} = \frac{\gamma f K}{mr}(r - f).$$

3.1.2 Interactive

The rate of change of herbivore biomass depends on plant biomass and the rate of change of plant biomass depends on herbivore biomass (negative feedback). A simple model for this situation is the Rosenzweig–MacArthur model:

$$\frac{dP}{dt} = rP\left(1 - \frac{P}{K}\right) - \frac{fPH}{1 + hfP} \tag{3.2a}$$

$$\frac{dH}{dt} = H\left(\frac{\gamma f P}{1 + hfP} - m\right). \tag{3.2b}$$

In this case, the loss rate of biomass from the plant, in the second term of (3.2a), depends on the herbivore biomass density, which is multiplied by the functional response of the plant, which in this case is the Holling type 2 functional response derived in Chapter 2. Therefore, the herbivore has a feedback effect on plant growth, which can create complex dynamics. The models considered in this book will all be of the interactive type.

The central theme of this book is the role that plant defenses, primarily chemical defenses, play in limiting the feedback effect of herbivores on plants in plant-herbivore interactions. However, much research has focused on other general ideas concerning the interaction of plants with herbivores and the rest of the food web. One is that plant quality, or the concentrations of nutrients in plants, limits herbivores. This limitation results from the fact that concentration of vital nutrients such as nitrogen and phosphorus are much lower in plant biomass than in the biomass of their herbivore consumers, so the consumer populations are required to find and digest large amounts of plant biomass to obtain sufficient nutrients to grow and reproduce. Another idea is that neither plant defenses nor quality are the main limiters of consumer populations, but the herbivores are limited by their own natural enemies: carnivores, parasitoids, viruses, and so forth. In this chapter we consider some of the central models of plant-herbivore interactions in which chemical defenses do not play an explicit role. These interactions include the top-down effects of herbivores on plants and the paradox of enrichment, both of which are caused by strong top-down effects of herbivores on plants. Also considered are the effects of natural enemies on the herbivore, and the phenomenon of insect herbivore outbreak, when the natural enemy can no longer control the herbivore. Finally, a stoichiometric model is presented that takes into account the nutrient concentration in a plant, and its effects on herbivore growth and survival.

3.2 The theory of top-down control

As we have seen in Chapter 2, and can be found from examining equation (3.2b), the non-trivial equilibrium from these equations holds the prey, or vegetation in this case, to the level

$$P^* = \frac{m}{f(\gamma - mh)}. \tag{3.3}$$

Of course, existence of this positive equilibrium requires that $\gamma - mh > 0$. Thus, according to this simple model of plant-herbivore interaction, the vegetation biomass is controlled to a level determined by the herbivore's parameters, and does not respond to changes in its own growth rate, r, or carrying capacity K. This is termed top-down control.

The importance of top-down control was recognized early in the second half of the twentieth century. The emphasis during the early days in ecology was on the effect of energy flow from the bottom-up on trophic structure. The trophic pyramid was assumed limited by energy transfer between trophic levels. Lindeman [221] assumed that the number of trophic levels was limited by the drastic losses of energy from one trophic level to the next as one goes up the food chain. But the paper by Hairston et al. [137] raised a very different perspective. They suggested that what was perhaps equally important to energy flow from the lower levels affecting the higher trophic levels was that predator-prey interactions exerted a control on lower trophic levels and affected the trophic structure and energy flow. Thus, rather than being passive recipients of energy from lower trophic levels, the higher trophic levels exert downward cascading effect, or top-down control. The Hairston et al. [137] concept of "top-down control" extended beyond simply herbivores and plants to longer food chains, usually ending with a first-order carnivore level in terrestrial food chains. Hairston et al. [137] believed that the herbivore level is kept low by predation from the first-order carnivore. This is represented in Figure 3.1. In aquatic pelagic systems, either first-order carnivores keep herbivores in check or second-order predators keep the first-order carnivores in check, as shown in Figure 3.2. The overall hypothesis, applied to all possible ecosystems, is sometimes called the ecosystem exploitation hypothesis [274], and implies that trophic efficiencies are determined by the predator-prey and competitive interactions and may be much more complex than what we have seen before.

There is an abundance of natural enemies of insect herbivores. For example, more than 60 different species of animals, including 30 species of birds, attack the Douglas-fir tussock moth. Evidence indicates that predators can sometimes regulate herbivore populations. Of the scale insects in a heavily infested pine stand, 97% were killed by a single species of predatory beetle in a 4-week period [291].

Debate exists about whether natural enemies can control herbivorous insects when their numbers are expanding. Parasitoids and predators respond to herbivorous insects in two ways, functionally and numerically. An individual natural enemy responds through a functional response, by increasing its rate of feeding on the prey species when its numbers increase. Of course, there is an upper limit on the rate at which a predator can feed or a parasitoid can lay eggs on its prey. Therefore, a functional response must "saturate"; that is, approach an asymptote as prey numbers increase. This asymptote is the maximum rate at which the predator can feed off the prey. This is incorporated into the functional response in a model of plant-herbivore-natural enemy, in which the natural enemy of the herbivore may be a carnivore or a parasitoid;

$$\frac{dP}{dt} = rP\left(1 - \frac{P}{K}\right) - \frac{f_1 PH}{1 + h_1 f_1 P} \tag{3.4a}$$

$$\frac{dH}{dt} = \frac{\gamma_1 f_1 PH}{1 + h_1 f_1 P} - \frac{f_2 HC}{1 + h_2 f_2 H} - m_1 H \tag{3.4b}$$

$$\frac{dC}{dt} = \frac{\gamma_2 f_2 HC}{1 + h_2 f_2 H} - m_2 C, \tag{3.4c}$$

where the functional response for the herbivore being fed on by the carnivore, C, is

$$\frac{f_2 H}{1 + h_2 f_2 H}.$$

To see that the carnivore regulates the herbivore biomass and that the plant biomass can increase with both K and r, solve equations (3.4a, b, c) at the positive equilibrium to obtain,

$$H^* = \frac{m_2}{f_2(\gamma_2 - h_2 m_2)} \tag{3.5a}$$

$$P^* = -\frac{1}{2}\left(K - \frac{1}{h_1 f_1}\right) + \frac{1}{2}\left[\left(K - \frac{1}{h_1 f_1}\right)^2 - \frac{4K H^*}{r h_1}\right]^{1/2} \tag{3.5b}$$

$$C^* = \frac{1}{f_2}\left(\frac{\gamma_1 f_1 P^*}{1 + h_1 f_1 P^*} - m_1\right)(1 + h_2 f_2 H^*). \tag{3.5c}$$

These equilibrium values are all positive for certain parameter ranges. In equation (3.5a) the herbivore is held to a value, usually low, determined by the carnivore's parameter values.

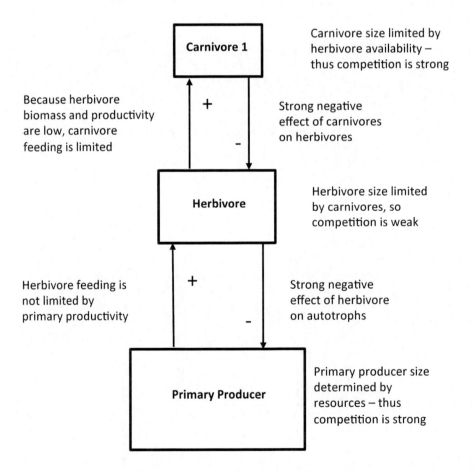

FIGURE 3.1: Three trophic level food chain, showing the effects of top-down control, as well as bottom-up effects.

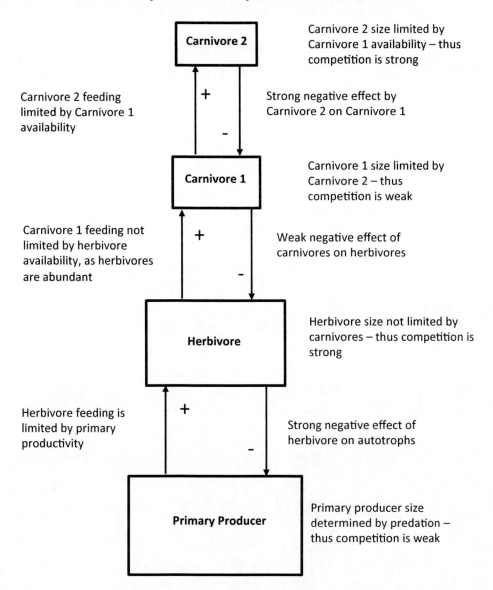

FIGURE 3.2: Four trophic level food chain, showing the effects of top-down control, as well as bottom-up effects.

However, this holding of the herbivore to a low level is dependent on the carnivore, or parasitoid, being able to increase numerically fast enough to keep up with the herbivore. Predators such as birds, small mammals, and many carnivorous insects, are unlikely to be able to keep up numerically with "outbreak" herbivore insects, that is, ones that can achieve very high growth rate under some environmental circumstances. But parasitoids, or infectious diseases could. The increase in numbers of predators due to an increase in prey numbers is called a numerical response, in contrast with the functional response, which refers only to the increase in rate of predation in response to prey numbers. Empirical evidence and modeling suggest that population cycles of some outbreak insects are controlled by internal agents, such as viruses. There are two lines of evidence that parasitoids, at least, often provide strong control of herbivorous insects, and predators sometimes do:

1. Outbreaks of the herbivorous insects often occur when parasitoids and/or predators are removed. This often happens when insecticides decimate carnivorous insects.

2. High destructivity of plants by herbivores often occurs when they are introduced to new areas in which they don't have natural enemies. That destruction is often brought under control when a natural enemy is introduced (biological control). An example is the larch casebearer, from Europe, becoming a pest when introduced to North America. The release of parasitic wasps from the casebearers' native locale brought the pest under control.

We will return to insect outbreaks later.

3.3 Paradox of enrichment

As discussed above, dynamic systems such as (3.2a,b) and (3.4a,b,c) often have at least one equilibrium in which all species have positive values for certain parameter ranges. However, sometimes this equilibrium is unstable, which again depends on specific parameter values. The conditions were derived in Chapter 2 (see Section 2.5) for stability of the Rosenzweig–MacArthur equations, also shown in (3.2a,b). It was shown that a small perturbation away from the equilibrium point could grow through time. In that case, instead of the predator and prey stabilizing to fixed values, the predator and prey cycle indefinitely.

We will go into more detail here, because this instability is well known in ecology as the paradox of enrichment, because, paradoxically, enriching the system by increasing the prey's carrying capacity leads to instability. In particular, changing the values of the parameters gradually can result in the system going through a Hopf bifurcation and changing from a stable state to an oscillatory one. For example, increasing the carrying capacity, K, leads to a transition from a stable steady-state equilibrium to limit cycle oscillations. This is the case, as it was shown in Chapter 2 (see (2.12) in Section 2.5) that it is necessary for instability that the plant equilibrium satisfy the inequality,

$$P^* < \frac{K}{2} - \frac{1}{2fh},$$

where

$$P^* = \frac{m}{f(\gamma - mh)}.$$

It can further be shown that

$$\frac{K}{2} - \frac{1}{2hf} = P_{peak}, \tag{3.6}$$

where P_{peak} is the peak of the humped curve in Figure 3.3. This can be shown by solving for the maximum of the plant isocline, which forms the hump, by setting the right-hand side of equation (3.2a) to zero;

$$H = \frac{r}{f}(1 + hfP)\left(1 - \frac{P}{K}\right). \tag{3.7}$$

We obtain (3.6) by differentiating (3.7) with respect to P, and setting the result to zero to obtain the maximum point.

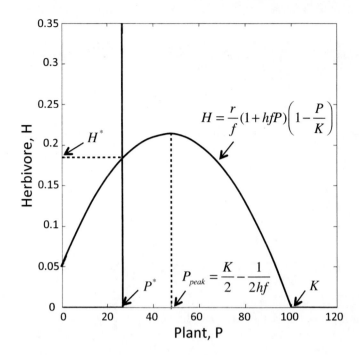

FIGURE 3.3: Zero isoclines of equations (3.2a,b), showing the equilibrium point formed by the intersection of the two zero isoclines being to the left of the peak of the prey zero isocline, and thus unstable. The parameter values are $r = 0.05. K = 100, f = 0.3, h = 0.5, m = 0.8$, and $\gamma = 0.5$.

The basic cause of oscillations in a system is usually something like this. A biological variable in the system has a tendency to grow in size, due to some self-amplifying, or positive feedback, effect. But a negative, or inhibitory, force develops that pushes the variable down again. This inhibitory force may be somewhat delayed, so the variable increases greatly before the inhibition starts, but then the inhibition becomes very strong and the variable may fall back to a lower level than it started at; that is, it may first "overshoot" due to the delay in the inhibitor, and then "undershoot" when the inhibitor has finally built up. The cycle repeats with the variable increasing again to "overshoot" and then suffers another "undershoot." This can continue indefinitely, creating periodic behavior.

An important part of the pair of equations (3.2a,b) is the term for consumption which has P in both the numerator and denominator

$$\frac{fPH}{1 + hfP}. \tag{3.8}$$

The coefficient of H in this term, representing per capita consumption by the herbivore, or "functional response," is a Holling type 2 functional response (HT2FR), which is identical to the Monod function or Michaelis–Menten function of biology and chemistry. Next to the logistic equation, this function is perhaps the most important in biology. In cell biology, it describes enzyme-mediated reactions. In ecology, it represents a feeding function that has saturation. By saturation is meant that the predator can consume only so much prey per unit time, no matter how large P is. If P approaches infinity, the feeding rate (3.8) approaches H/h, that is, the maximum feeding rate of an individual predator, $1/h$, multiplied by the herbivore biomass. The fact that the herbivores can eat only so fast means that when

P reaches a very high level, the plant biomass is for a time virtually uncontrolled by the herbivore, and the plant population growth slows down only due to approaching its carrying capacity, K. But the herbivore population by that time has grown to a large size and knocks the plant population down to a very low level. The herbivore population then crashes and the cycle starts over. The HT2FR is why equilibrium points to the left of the peak are unstable. Any slight decrease in P from the equilibrium point will be amplified, because the per capita prey mortality increases.

It is useful to study this system in detail. Recall the positive steady-state equilibrium is

$$P^* = \frac{m}{\gamma f - fhm}$$

$$H^* = \frac{r}{f}(1 + fhP^*)\left(1 - \frac{P^*}{K}\right).$$

(3.9)

As a start to understanding what makes the system stable or unstable, it helps to look at Figure 3.4, where the signs of dP/dt and dH/dt (derivatives now represented by dots) are shown in each of the four regions divided by the zero isoclines. Why are equilibria to the left of the peak unstable, whereas to the right of the peak they are stable? Let's look at a close-up of the equilibrium when it is to the left of the peak (Figure 3.5). When the numbers of herbivores and plant biomass are right on the equilibrium point, there is no tendency to change, because $dP/dt = 0$ and $dH/dt = 0$. However, suppose we perturb prey numbers, P away from the equilibrium point. If we perturb P to a lower value, and do not change H, the system moves into a region where $dP/dt < 0$. That is important, because usually we think of predation on a population going down when its numbers decrease, something called "compensation." However, in this case, prey per capita predation increases, something called "depensation." This is entirely due to the HT2 functional response (the coefficient of H in equation (3.8)). When P decreases, the denominator of the HT2FR decreases, so that the prey per capita predation increases. Biologically, this happens because a decrease in prey biomass, P, causes the predators to be less "saturated" with prey, so they spend more time hunting for food. The result is that an initial decrease in P is reinforced by further decrease. Note that in this region, $dH/dt < 0$, so the predator is also decreasing. The net result is that the initial perturbation is pushed farther from the equilibrium.

If we carefully construct the trajectory that results from our perturbation, or from a perturbation in any direction, we would see that the trajectory moves away from the equilibrium, but in an outwardly counter-clockwise direction (Figure 3.6). Figure 3.6a shows the trajectory of plant and herbivore spiraling outward from the unstable equilibrium, while Figure 3.6b shows the plant and herbivore biomasses cycling through time, with the herbivore lagging by 90°. Figure 3.6c shows the trajectory approaching a limit cycle.

What is most fascinating about this phenomenon of instability is that the shift from stable to unstable equilibrium occurs as the carrying capacity, K, is increased. It is easy to see from the equation (3.6) that increasing K shifts the peak (P_{peak}) of the humped prey isocline (equation (3.7)) to the right. It seems paradoxical that an increase in the carrying capacity of the plant, which would appear to be valuable to an ecosystem, at least in terms of allowing it to reach higher biomass, actually causes the system to become unstable, at least in this simple model of plant and herbivore. Rosenzweig [316], who discovered this instability, a Hopf bifurcation, referred to it as the "paradox of enrichment."

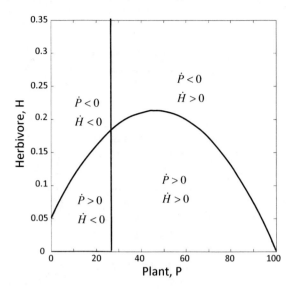

FIGURE 3.4: Zero isoclines of equations (3.4a,b), showing the equilibrium point formed by the intersection of the two zero isoclines, and the signs of the right-hand sides of these equations in each of the four regions formed by the intersection of the zero isoclines. Parameter values same as in Figure 3.3.

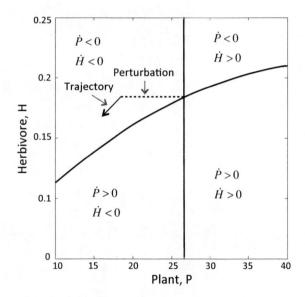

FIGURE 3.5: Close-up of state plane with equilibrium point left of peak. The dashed line represents a perturbation that decreases P to a value lower than its equilibrium value. The arrow shows the initial direction of the perturbed trajectory.

3.4   Collapse of a grazer-plant system

It was noted above that the plant-herbivore system could have at least one equilibrium point in which both species have positive values. Actually the system described by equations

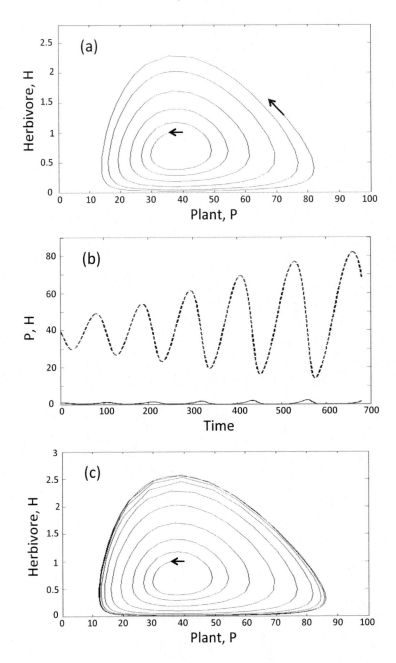

FIGURE 3.6: (a) Initial limit cycle oscillations following a perturbation of the plant-herbivore system described by equations (3.2a,b) with parameter values of Figure 3.3. (b) Initial oscillations of plant, P (dashed line), and herbivore, H (solid line), biomass. (c) Continuing spiraling toward the limit cycle.

in (3.2) can have only one such "interior" equilibrium, which is clear from the fact that the vertical herbivore zero isocline can intersect the prey zero isocline only once. However, other forms of the zero isoclines are also possible and can affect the number and nature of the equilibria. This can have strong connections to important ecological phenomena, including human-managed systems. A grazing system ecologist, Noy–Meir [268], utilized the consumer-plant (grazer-grass) model to understand the sharp sudden tipping that seemed possible in such systems.

Let us start with the familiar prey-predator equations, with plants (P) and herbivores (H) represented as

$$\frac{dP}{dt} = r_1 P\left(1 - \frac{P}{K}\right) - c_1 H\left(\frac{P}{1 + hP}\right) \tag{3.10a}$$

$$\frac{dH}{dt} = H\left(-a + c_2\left(\frac{P}{1 + hP}\right)\right). \tag{3.10b}$$

We use here the form of the equations and parameters used by Noy–Meir [268], except that we retain P to represent plant biomass, whereas Noy–Meir used V for plant, or vegetation, biomass. These equations would produce isoclines that were shown in Figures 3.3, 3.4, and 3.5. However, Noy–Meir realized that human-controlled grazing systems work in a different way than the equations for freely interacting plant and herbivore populations shown above, as humans can exert some level of control, as in grazing systems. In particular, the herbivore (sheep in this case) number (or biomass) can be set at some value $H = H_0$, rather than varying freely. This represents the fact that humans control the sheep density. Doing this in the above equations, we can look at the vegetation equation by itself; that is, just equation (3.10a)

$$\frac{dP}{dt} = r_1 P\left(1 - \frac{P}{P_{max}}\right) - c_1 H_0\left(\frac{P}{P + k}\right) \tag{3.11a}$$

which Noy–Meir wrote as

$$\frac{dP}{dt} = G - C, \quad G \equiv r_1 P\left(1 - \frac{P}{P_{max}}\right), \quad C \equiv c_1 H_0\left(\frac{P}{P + k}\right). \tag{3.11b}$$

The expression for G above is simply logistic growth of the plant. The expression for C, or consumption by the herbivore has a saturating functional response. Structurally, the equation for the plant biomass is the same as equation (3.2a), but the herbivore density is fixed at an imposed level, H_0. Now one can find the equilibria, using a plane, though it is not the state-plane as before, but simply a plot of $G(P)$ and $C(P)$ against P.

Note that different values of H_0 (or just call them constant H) produce different $C(P)$ curves in Figure 3.7a. Each of these curves represents a different stocking density of sheep. These curves of $C(P)$ could be plotted along with the $G(P)$ curve (Figure 3.7b). The intersections of the two curves represent the vegetation equilibria. Note that for different curves, $C(P)$, there can be different numbers of equilibrium points. For low values of sheep stocking density H_0, the $C(P)$ curves only intersect values of the right-hand side of $G(P)$, corresponding to large values of equilibrium plant biomass, P^*. For sufficiently large values of H_0, e.g., $H_0 = 3.0$, there can be two equilibrium points, while for still larger values of H_0, $H_0 = 4.0$, there are no equilibrium points.

Now consider the dynamics associated with each type of situation. For low values of H_0 (Figure 3.8a, with $H_0 = 1.0$) the one equilibrium point is stable, as vegetation tends to decrease when it is higher than this point and to decrease when it is lower. In Figure 3.8b, when H_0 has increased to 2.0, a new equilibrium for a smaller value of P^* has emerged. This point is unstable, as plant biomass increases when perturbed slightly above this value and decreases when perturbed slightly below. In the former case, the plant biomass will grow toward the equilibrium point for larger P^*, while in the latter case it will decrease toward zero. The same dynamics hold in Figure 3.8c, except that here these two equilibria are converging toward each other. When H_0 is increased only slightly further, something

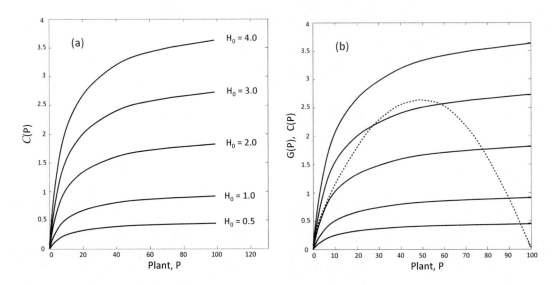

FIGURE 3.7: (a) Plots of $C(P) = c_1 H_0(P/(P + k))$ versus P for different values of H_0. (b) Plots of $C(P)$ superimposed on $G(P) = r_1 P(1 - P/P_{max})$. Parameter values are $c_1 = 1, k = 10, P_{max} = 100, r_1 = 0.05$.

new happens: both equilibria disappear (Figure 3.8d). At this point, the only dynamics remaining for the plant biomass is to decrease toward zero.

The ecological importance of this analysis is that it shows the possibility of complete collapse of a plant-herbivore system when herbivory, in this case grazing, is imposed without any feedback from the plant biomass on the herbivory. When there is feedback; that is, when equation (3.10b) as well as (3.10a) act, then feedback from declining plant numbers would feed back on herbivory to slow it down. But in this case, nothing stops the gradual increase in herbivory until the plant positive equilibrium disappears.

3.5   Herbivore outbreaks

A small percentage of herbivorous insect species are "outbreak species," which escape regulation by natural enemies and grow rapidly into large populations at times and can defoliate tree species. This is true even in tropical forests, though most severe outbreaks occur at high latitudes. Outbreaks can occur rapidly and may lead to 50% foliage consumption within 2 to 3 years. A typical outbreak species is the Douglas-fir tussock moth. Before an outbreak in Arizona, there was one larva per 6250 cm^2 of foliage in Arizona stand. At peak, 2 years later, there were 15 larvae per 6250 cm^2. In another case, in the Engadine Valley of Switzerland, larch budmoth increased an average of 18,000-fold in a 1- to 2-year period [291].

Specialists in insect outbreaks classify four phases of outbreak: Release phase in which natural mechanisms no longer control the insect; a peak phase reached by exponential increase; a decline phase, which can be an extremely sharp drop; and a post-decline phase of continuing slow decrease. The release phase is crucial. The insect climbs beyond a threshold

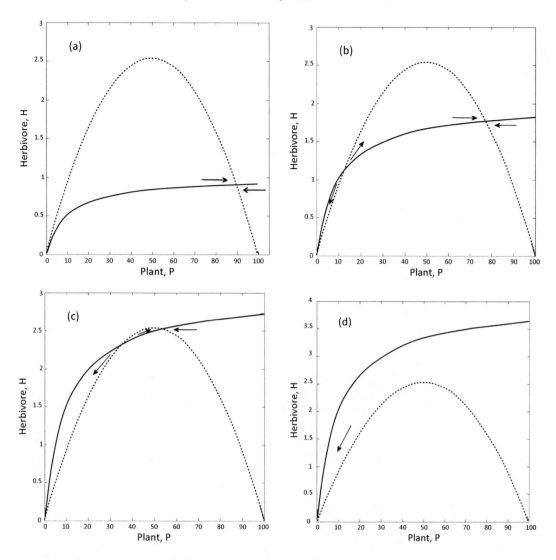

FIGURE 3.8: Plot of $G(P) = r_1 P(1 - P/P_{max})$ with $C(P) = c_1 H_0 P/(P + k)$ for (a) $H_0 = 1$, (b) $H_0 = 2$, (c) $H_0 = 3$, (d) $H_0 = 4$.

where natural controls can no longer control them. This is clearly something that foresters and ecologists alike want to understand.

What are the causes of these outbreaks? Environmental factors can affect outbreaks, and outbreaks of some species of insects have increased in recent times; perhaps due to logging practices and fire suppression, leading to greater densities of tree species preferred by particular insects; e.g., there were nine outbreaks of eastern spruce budworm during the nineteenth century, whereas during the first 80 years of the 20th century there were 21 (and more widespread) [291]. Pesticides may also contribute by killing biological control agents (carnivores like spiders and predatory beetles), which are slower to recover than the pests themselves. The main factors seem to be either environmental conditions, such as a favorable food supply, that allow the herbivorous insects to grow too fast for the usual control agents to keep up with them, or something that hinders these control agents.

Consider an example described by White [383]: the spruce budworm, which feeds preferentially on balsam fir. The caterpillars of this moth do not feed immediately after emergence from the egg, but are dispersed by wind and hibernate over winter. When they emerge in spring they start to feed on *old* fir needles; that is, they are senescence feeders, feeding on needles that are breaking down and releasing nutrients, i.e., amino acids, in high concentrations during nutrient translocation. Outbreaks tend to occur when there are large numbers of old trees, when it is likely that many of the needles are old and not very vigorous, so they tend to break down quickly. But the presence of a large stand of old trees is not sufficient in itself to lead to a spruce budworm outbreak. The period of 1948 to 1958 in eastern Canada was a period of summer droughts interspersed with wetter than normal winters. This combination of drying of roots in the summer and waterlogging them in the winter caused dieback of crowns, with rapid aging of leaves and release of high concentrations of amino acids being translocated out of the dying leaves. The extra amino acid in the diets of the spruce budworm caterpillars was enough to increase caterpillar survival, which led to population explosions that predators and parasitoids could not control. DDT was sprayed to stop the outbreaks, but that just made things worse by saving the old stands of trees and letting them get older and even more susceptible to outbreaks.

As in the case of grazing system collapse, the phenomenon of an insect outbreak can be described as a mathematical "catastrophe" or "critical transitions." This was recognized by Ludwig Jones, and Holling [232], who developed a model for the process, which involved, in their case, escape from natural enemies.

In the model of the collapse of the grazing system, the effect of the herbivore was gradually increased, leading to sudden overexploitation of the vegetation, so it was the equation for the vegetation that was the focus of the model. In the case of herbivore outbreak, it is gradual increase in the vegetation that leads to a threshold at which the herbivore is sufficiently hidden from predators that it can suddenly explode in population. Ludwig et al. [232] modeled the spruce budworm, which feeds preferentially on balsam fir, but also on other conifers in the boreal forest. We will use the notation of Ludwig et al. [232].

The equation for spruce budworm density was hypothesized to have the following form

$$\frac{dB}{dt} = r_B B \left(1 - \frac{B}{K_B}\right) - \beta \frac{B^2}{\alpha^2 + B^2}. \qquad (3.12)$$

As in the case of grazing system collapse, only one equation is used to describe the dynamics, only in this case it is the equation for the herbivore. The plant, the spruce vegetation, is assumed to simply grow with no feedback effects from the budworm in the budworm outbreak phase, which is described by the model. The first two terms represent logistic growth, with a maximum growth rate r_B and carrying capacity K_B. The logistic term is straightforward. The budworm population, if not regulated by natural enemies can grow to the level sustainable by spruce foliage, K_B. The second term represents predation by natural enemies, parasitoids and birds, on the budworm, as β is a measure of strength of that effect. Note that the functional response used here is a Holling type 3 response (see Chapter 2).

What is meant by $G(B) \equiv \beta B^2/(\alpha^2 + B^2)$? The Holling type 3 signifies a search image phenomenon; that is, the rate of feeding grows faster than the population of the prey, because the predator's search image, or skill at finding the prey increases with the number of times it encounters prey individuals. As in the HT2FR, saturation of the natural enemies when B is large. β is important because it will gradually decrease as spruce foliage increases through time, following the previous defoliation by budworms; i.e., budworm protection from natural enemies will gradually increase with foliage, as it is difficult for the natural enemies to detect the budworms in dense foliage. Ludwig et al. [232] rescaled the equation, by letting

$$\mu = \frac{B}{\alpha}, \quad R = \frac{\alpha r_B}{\beta}, \quad Q = \frac{K_B}{\alpha}.$$

They rewrote the equation with new variables and set the right-hand side to zero to yield, at equilibrium,

$$R\left(1 - \frac{\mu}{Q}\right) = \frac{\mu}{1 + \mu^2}. \tag{3.13}$$

This is equivalent to saying that the "fast" variable, B, comes to equilibrium rapidly with the changes in parameters R and Q. The parameters R and Q are actually slowly changing as forest conditions change. In particular the increasing density of spruce foliage decreases the ability of natural enemies to find and attack the budworms, so β would be likely to decrease through time and α to increase as foliage increases from the last defoliation event. Thus both $R = \alpha r_B/\beta$ and $Q = K_B/\alpha$ would be slowly increasing through time.

As in the case of the example of grazing system collapse described above [268], we can find the equilibria by plotting both sides of the equation versus B (rescaled as μ). The left-hand side gives a straight line with y-axis intercept of R, and slope $-1/Q$, while the right-hand side produces a humped curve (Figure 3.9a). What is interesting is that again we can have up to three roots or equilibria; where B_- and B_+ are stable points and B_c is unstable. Assume the budworm density is initially low (at B_- in Figure 3.9a). As R increases, B_- and B_c converge and eventually disappear, leaving B_+ as the only equilibrium, to which the budworm population rapidly grows (Figure 3.9b).

Then Ludwig et al. [232] use a bifurcation diagram (Figure 3.10), with the parameters Q and R as the x- and y-axes, respectively, to illustrate the phenomenon. The two-branched curve in the figure divides the plane into three regions. Below the lower curve (small R and Q), only one root exists, that with the smaller value of μ. As Q increases, with R still relatively small, a region is reached between the two branches, where all three roots occur. This is the bistable range, where the budworm could be at either high or low equilibrium

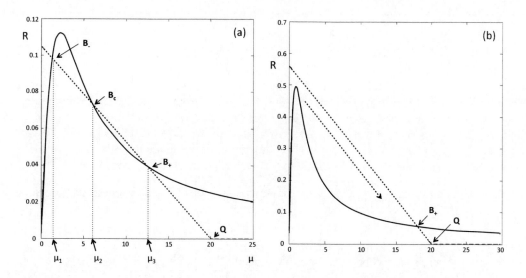

FIGURE 3.9: Plots of the right-hand side (solid line) and left-hand side (dashed line) of equation (3.13) as functions of μ. (a) For $Q = 20$ and $R = 0.105$, three intersections of the curves are shown, for three equilibrium points. B_- and B_+ are stable, while B_c is unstable. (b) For $Q = 20$ and $R = 0.55$, only equilibrium point B_+ exists.

density. As R increases, so that we are above the upper branch, only the equilibrium with the budworm at high density can occur.

The model of Ludwig et al. [232] shows that at least some insect outbreaks can be described from the mathematical point of view as parameters reaching critical thresholds. This unites these outbreaks with a wide spectrum of phenomena in nature that can be described by critical transitions [322].

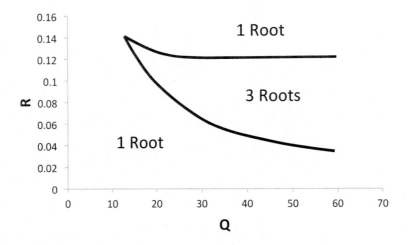

FIGURE 3.10: Bifurcation diagram showing the location in (R, Q)-space of the locations of regions where either three roots or only one root exist.

3.6 Effects of plant quality

It is known that the "quality" of plant material plays a major role in the interaction between the plant and herbivore. Plant tissue is composed mostly of cellulose and lignin (carbohydrates), but insect and vertebrate tissue is mostly protein and fat. Because nutrient concentrations in mature foliage rarely exceed 2% in conifers and 3% in deciduous trees, whereas the bodies of herbivores such as insect larvae contain more than 10% nitrogen (N) it is likely that herbivores are limited by nutrients rather than energy content of food. Therefore, the nutrient:carbon ratio in plant biomass is important. It affects the efficiency of conversion of ingested food, that is, the assimilation efficiency, which varies with nitrogen and other nutrient content. This efficiency of conversion is less than 20% for invertebrate herbivores feeding on most tree tissue, compared with 50% for many invertebrate carnivores. For example, xylem sap-sucking insects (e.g., aphids, aldelgids) ingest 100 to 1000 times and chewing arthropods 5 to 6 times their body weight each day [291].

Besides protein, some other nutrients may be scarce in plants (methionine, water-soluble B vitamins, sodium) and may be limiting for some herbivores. T. C. R. White, in a book called *The Inadequate Environment* [384], had the view that the primary factor limiting herbivore populations relative to their apparent food supply is the shortage of nitrogen (N) in their diets. Outbreaks, in his view, are triggered by increased levels of soluble N in plant tissue during droughts, as we mentioned earlier. White's hypothesis is sometimes called the "green desert" hypothesis; i.e., the environment of a forest looks green, but there's really not as much to eat that is nutritious as one might think. The problem is that the N is fairly finely spread, even in leaves, and it is often tied up in complex structural material. The herbivore has to spend both time and energy obtaining that nitrogen, all the while having to make up for its own bodily losses. So it's really not a matter of absolute amounts but rather of the economics of time and energy of getting what the herbivore needs.

Some models take into account the effects of cycling of a limited amount of nutrient in the ecosystem and its effects on the plant-herbivore interaction. The model of Loladze et al. [229] assumes that nutrient limitation can affect both the carrying capacity of the plant and the feeding efficiency of the herbivore. Again, we will primarily use the notation used by the authors. Loladze et al. introduced a variation on the plant-herbivore interaction model that incorporates effect of phosphorus as a limiting nutrient;

$$\frac{dP}{dt} = bP\Big(1 - \frac{1}{\min(K/P, \eta/q)}\Big) - \frac{cPH}{\alpha + P} \tag{3.14a}$$

$$\frac{dH}{dt} = e\min\Big(1, \frac{\eta}{\theta}\Big)\frac{cPH}{\alpha + P} - dH. \tag{3.14b}$$

Here

$\eta(t)$ = nutrient per unit plant carbon (variable),

θ = nutrient per unit herbivore carbon (fixed),

q = minimum plant nutrient to carbon ratio.

The $\min(K/P, \eta/q)$ term means that the carrying capacity is determined by whatever is limiting, either by carbon, in which case it is K, or by phosphorus, in which case the carrying capacity is q, which is the minimum fraction of phosphorus in the vegetation for which it is viable.

The $\min(1, \eta/\theta)$ term means that the assimilation efficiency is 1 if the ratio of the fraction of phosphorus to carbon in the plant's edible biomass is equal to or greater than the fraction in herbivore biomass, or is equal to η/q otherwise.

Thus phosphorus limitation may affect both the carrying capacity of the vegetation and the efficiency to the herbivore. Total phosphorus, P, is conserved (closed system is assumed regarding nutrient); i.e., P = constant, so that

$$P = \eta(t)P(t) + \theta H(t) \quad \text{or} \quad \eta(t) = \frac{P - \theta H(t)}{P(t)} \tag{3.15}$$

where $\theta H(t)$ is the amount of phosphorus stored in herbivore biomass and $\eta(t)P(t)$ (where $\eta(t)$ is variable) is the amount of phosphorus stored in the plants.

This model, along with a variation by Kuang et al. [205], has been simulated as a function of increasing light, as shown in Figure 3.11. The four panels in Figure 3.11 show simulations in time of equations (3.14a,b) for four different levels of light availability, which determines K, the carrying capacity of plant if carbon is limiting.

Case A. Under low light conditions, the plant is limited by carbon, so its P:C ratio is high, but plant biomass is low enough that herbivores are maintained at fairly low levels (Figure 3.11a).

Case B. When light and thus K is increased, the plant-herbivore interaction becomes unstable, due to the paradox of enrichment (Figure 3.11b).

Case C. Further increase in K leads back to a stable system with high biomass in both the plant and herbivore, but both the plant and herbivore reach much higher biomasses than in Case A (Figure 3.11c).

Case D. As K is increased still further, herbivore biomass steadily decreases, because the nutrient per plant biomass, η, decreases, so the herbivore can assimilate less and less of the biomass.

In Figure 3.12 these transitions are expressed in a bifurcation diagram, showing a Hopf bifurcation, then the effects of decreasing plant quality, nutrient dilution, on the herbivore.

The results of Loladze et al. [229] show the importance of plant quality on the interaction with an herbivore consumer. When plants are growing under high light and low nutrient conditions, the herbivore is dependent not on the energy content of the plant biomass, but the nutrient content. High light conditions can cause the nutrients in plant material to be

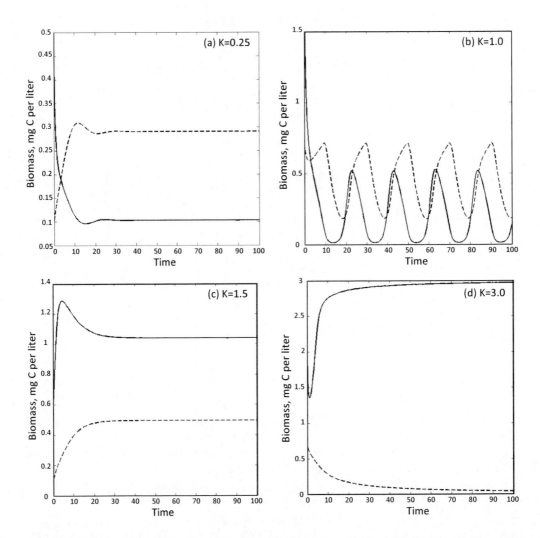

FIGURE 3.11: Plots of plant (solid line) and herbivore (dashed line) biomass from simulations of equations (3.14a,b) for various levels of plant carrying capacity, K. (a) $K = 0.25$, (b) $K = 1.0$, (c) $K = 1.5$, (d) $K = 3.0$. Other parameter values are $N = 0.025$, $e = 0.8, b = 1.2, d = 0.25, \theta = 0.3, q = 0.0038, c = 0.81, \alpha = 0.25$.

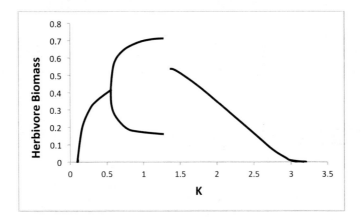

FIGURE 3.12: Bifurcation diagram showing the changes in the herbivore biomass as the carrying capacity determined by carbon, K, is increased. Initially, herbivore biomass grows rapidly with K, then becomes unstable with the amplitude indicated, and then declines with further increase in K.

diluted so that, while the herbivore's energy needs may be met, its nutrient needs are not, so the herbivore declines.

This model illustrates the problems that herbivores face in exploiting plants, where the nutrient content can become so low that herbivore growth and reproduction is inhibited. Herbivores have many ways of circumventing nutrient deficiency, such as by targeting the highest quality plant biomass available. Many specific techniques are described by White [383], and some are noted in Chapter 6.

3.7 Glossary

Bifurcation diagram: This is a diagram showing the changes in the steady-state solutions and stability of a system as some control parameter is changed across some range of value.

Bottom-up control: This term is commonly applied to the concept of the size of populations or trophic levels being controlled by their resources.

Compensation: This happens in a population when a decrease in the standing stock of the population leads to an increase in the growth or survival rate of the population.

Critical transitions: A critical transition can occur when a system has two or more alternative attractors and a change in external conditions move the system past a threshold point at which it inevitably switches from its original state to a new one. The collapse of the grazing system is an example in which increased grazing pressure, which, at some point, called a tipping point, causes the system to collapse. This is also called a "catastrophe," from mathematical catastrophe theory.

Depensation: This happens in a population when a decrease in the standing stock of the population leads to a decrease in the growth or survival rate of the population.

Ecosystem exploitation hypothesis (or EEH): The EEH hypothesizes that the number of trophic levels is determined by the level of primary production. In tritrophic food changes (plant-herbivore-carnivore) it also assumes that the herbivore trophic level is controlled by the carnivore trophic level.

Functional response: The rate of intake of prey numbers or biomass per predator or per unit predator biomass. The Holling type 2 functional response is an example.

Green desert hypothesis: This hypothesizes that the generally low nutrient content of plant biomass limits the population sizes of herbivores, even when the abundance of plant biomass is high.

Hopf bifurcation: A Hopf bifurcation occurs when a parameter is changed such that the system changes from one that has local stability to one that has periodic limit cycles.

Interactive model: An interactive model of a consumer and resource (a herbivore and plant as an example) is one in which the consumer has a negative feedback effect on the growth rate of the resource.

Limit cycle oscillations: A closed curve in the consumer-resource (e.g., plant-herbivore) state plane that originates in a nonlinear consumer-resource model when a stable equilibrium point becomes unstable through a Hopf bifurcation. The trajectory spirals out from the unstable point and approaches the points of the limit cycle asymptotically.

Non-interactive model: A non-interactive model of a consumer and resource (an herbivore and plant as an example) is one in which the consumer has no effect on the growth rate of the resource.

Outbreak herbivores: These are herbivores that tend under favorable circumstances to explode into population sizes orders of magnitude greater than the usual size.

Paradox of enrichment: This refers to the possibility, in consumer-resource models such as the Rosenzweig–MacArthur model, for a gradual increase in the carrying capacity of the resource to suddenly lead a stable equilibrium to turn into limit cycle oscillations, by undergoing a Hopf birfurcation.

Plant quality: Plant quality usually refers to the concentrations of nutrients such as nitrogen and phosphorus in parts of the plant that are eaten by herbivores

Regime shift: A regime shift is a sudden jump of a system from one state to another. A critical transition is a special case of a regime shift in which two alternative attractors exist.

Stoichiometry: Stoichiometry refers to the balance of energy and the various bioelements that are utilized by organisms and occur in organism biomass in particular proportions, which, while typically relatively fixed, may also vary depending on circumstances.

Top-down control: This term is commonly applied to the concept of the size of populations or trophic levels being controlled by, respectively, populations or trophic level that feed on the population or trophic level in question.

Trophic level: The trophic level of a species is defined by its feeding relationships; in particular, how many energy transformation steps it is away from the basic source of energy, which is usually the sun. Thus a herbivore feeds on vegetation, which uses solar energy to build biomass, so its trophic level is two. The trophic level of a food chain may refer to all of the organisms that are the same number of energy transformation steps from the sun.

Chapter 4

Models with Toxin-Determined Functional Response

In this chapter, several models that explicitly incorporate plant toxins will be discussed. The first models described, toxin-determined functional response models (TDFRM), use modifications of the traditional functional responses (such as Holling types) by considering a reduced or non-increased rate of herbivore browsing on plant species that are more toxic. The models of this type are introduced from simple (involving only one plant species and one herbivore population) to more complex models (with multiple plant species as well as a predator population for the herbivore). These models focus on the herbivore responses to given toxicities of one or two plants. To validate the model with two plant species and one herbivore, solutions of the TDFRM are compared with a field experiment of the long-term ecological research (LTER) project involving willow (less toxic plant), alder (more toxic plant), and moose. The comparison also illustrates that the TDFRM can provide a much better fit to the observations than the corresponding model with Holling type 2 functional response. Detailed bifurcation analyses of the 2-dimensional TDFRM are also provided for various scenarios according to the existence of equilibrium points, periodic solutions, and their stabilities (local and global). Bifurcations include Hopf and homoclinic bifurcation. A reaction-diffusion model with the toxin-determined functional response is also discussed. For this model, the existence of traveling solutions is considered and the stability analysis provides conditions for the invasion of toxic plant species in space. The modeling studies described in this chapter include results presented in [53, 96, 97, 98, 218, 226, 227, 348, 389].

4.1   A simple TDFRM with a single plant population

As illustrated in the previous chapters, the most commonly used functional response is the Holling type 2 response, which is characterized by a monotonic increase in intake up to an available biomass beyond which the predator's ability to consume more biomass is exceeded. As in the Holling type 2 response, many other traditional functional response models specialized to mammalian browsing (see, for example, [17, 69, 155, 233, 234, 285, 298, 338]) do not include explicitly the effect of plant toxicity on plant-herbivore interactions. But many plants are heavily defended against herbivory by secondary chemicals, and in many mammal-plant interactions these toxins determine satiation (reviewed in [112], see also [42, 64, 83, 248, 282, 298, 307]). Research on herbivores indicates that aversion of toxic plants (conditioned food aversions) can occur in large domestic animals (e.g., [298]) and other herbivores (see, for example, [83, 241]). The implications for plant and herbivore population and community dynamics of plant defenses have been explored to only a small extent.

Plants defend themselves against consumption by herbivores through a variety of secondary chemicals that are toxic to herbivores, or decrease their ability to digest plant biomass. This has consequences for the dynamics of plant-herbivore interaction. Chemically mediated interactions between plants and herbivores have been shown to play an important role in ecology, evolutionary biology, and resource management (e.g., see [38, 44, 41, 70, 282, 315, 368]). Although recent research indicates that a large part of the explanation for why "the world is green" involves the top-down control of predators on herbivores (e.g., [351]), defensive chemicals clearly play a role both in directly limiting the amount of plant biomass consumed and in indirectly reducing it by inflicting higher mortality and lower growth and reproduction of herbivores (e.g., [261]).

One consequence of feeding on plants containing toxins is that rates of ingestion may be limited by a herbivore's ability to avoid toxins or detoxify food rather than to mechanically process food. Not surprisingly, herbivores have developed a host of physiological and behavioral mechanisms to deal with plant secondary metabolites [183]. Physiologically, vertebrates can regulate absorption of plant toxins by gut cells, respond to chemically mediated taste and trigeminal stimulation, and detoxify lipophilic compounds via enzymatic biotransformation [83]. For instance, marsupial folivores oxidize plant terpenes using P450 enzymes, and species with diets high in monoterpenes exhibit greater capacity for biotransformation of toxins than their generalist counterparts [33]. Behaviorally, vertebrates can select plants or plant parts containing low concentrations of a toxin [42], manage food to leach toxins from plants [84, 260], self-medicate to ameliorate effects of toxins [369], and adjust meal duration and intake per meal [241, 337]. An ability to regulate intake of plant secondary metabolites has been reported for several species of vertebrate herbivores [34, 171, 337]. For instance, brushtail possums ate more of the toxin benzoate when the rate at which it could be detoxified by conjugation was increased by adding glycine to the diet [240]. Herbivores also achieve greater intake of nutrients by selecting mixed diets containing foods processed by different detoxification pathways, thereby avoiding saturation of any particular pathway [242, 299].

Besides limiting plant consumption, chemical defenses may have implications for the paradox of enrichment [316], which predicts that increasing plant carrying capacity can lead to destabilization of plant-consumer interactions under certain conditions, leading to limit cycle oscillations. Limit cycles can arise in consumer-resource interactions with a Holling type 2 functional response [158, 159], because with that saturating response the per capita prey feeding rate of the consumer population decreases with increasing plant density. The effect of the toxicant could be similar, if in effect it imposes additional resource handling time on the herbivore. On the other hand, if toxicants can decrease the herbivore ability to graze plants to low levels, it is also intuitively possible that the likelihood of such limit cycles could be reduced.

Many other patterns of plant-herbivore interactions, at the population, community, and ecosystem levels, are affected by plant chemical defenses. For this reason, it is important to understand theoretically the sort of effects that can be expected. In such a model, the functional response needs to incorporate the occurrence of a toxicant in a plant that, above certain levels of intake by the herbivore, can have a negative effect on herbivore growth.

4.1.1 Derivation and emerging properties of the model

Some of the earliest plant-herbivore models that incorporate explicitly the effect of plant toxins on herbivore dynamics are formulated and studied in [97, 98, 226]. This type of model has been termed toxin-determined functional response model (or TDFRM). In this section, the simplest case of a 2-dimensional TDFRM is presented, which will be referred to as the 2-D TDFRM.

Let $P = P(t)$ and $H = H(t)$ denote the densities of plant and herbivore biomass at time t. The model reads

$$\frac{dP}{dt} = rP\left(1 - \frac{P}{K}\right) - C(P)H,$$
$$\frac{dH}{dt} = BC(P)H - dH, \tag{4.1}$$

where the functional response is described by

$$C(P) = f(P)\left(1 - \frac{f(P)}{4G}\right), \tag{4.2}$$

and $f(P)$ is the Holling type 2 functional response, i.e.,

$$f(P) = \frac{e\sigma P}{1 + he\sigma P}. \tag{4.3}$$

In equation (4.1), r is the intrinsic growth rate of the plant, K is the plant's carrying capacity, B is the conversion constant (herbivore biomass per unit of plant), and d is the per capita death rate of herbivore. The toxin-determined functional response $C(P)$ in equation (4.2) contains, as one factor, the traditional Holling type 2 functional response (4.3), representing ingestion per unit time derived on the basis of search by the consumer moving at a constant speed through space with randomly distributed prey of biomass density P. We will refer to this type of functional response that explicitly incorporates plant toxins as toxin-determined functional response, or TDFR. The parameter e in equation (4.3) is the resource encounter rate, which depends on the movement velocity of the consumer and its radius of detection of food items. The parameter σ ($0 < \sigma \leq 1$) is the fraction of food biomass encountered that the herbivore ingests, while h is the handling time per unit biomass of plant, which incorporates the time required for the digestive tract to handle the item. The second part of the toxin-determined functional response, that is, $1 - f(P)/(4G)$, accounts for the negative effect of toxin, where G represents the level of plant toxicity defined by $G = M/T$. The parameter M is a measure of the maximum amount of toxicant per unit time that the herbivore can tolerate (g toxin kg^{-1} herbivore biomass day^{-1}), and T is the amount of toxin per unit plant biomass (g toxin kg^{-1} plant biomass). Hence the units of G are [(g toxin kg^{-1} herbivore biomass tolerated day$^{-1})$ (g toxin kg^{-1} plant biomass)$^{-1}$]. G

decreases when the concentration of toxin tolerated by the herbivore decreases or the concentration of toxin in the plant biomass increases. Therefore, the smaller the value of G, the larger the effect the toxin has on intake. If M is assumed to be the same for all individuals within an herbivore population, then the variation in G will result from the different plant species encountered. The factor 4 is a multiplier that guarantees that the function $C(N)$ is nonnegative and does not exceed G.

When the effect of toxin is sufficiently high, that is, $G \ll 1$, it can be shown that $C(P)$ is approximately

$$C(P) \approx \frac{e\sigma P}{1 + he\sigma P + \frac{e\sigma P}{4G}}. \tag{4.4}$$

What this means is that in the limit of high toxin concentration, the effect of toxin is simply equivalent to an increase in handling time by the herbivore; that is, toxin slows down ingestion further than handling time alone. This type of simple increase in handling time would also occur from non-toxic digestion-slowing plant material, such as fiber. However, $C(P)$ was designed so that for smaller values of G (greater toxicity), there can be more severe consequences for an herbivore population. The parameter h, where $1/h$ is the maximum consumption rate by the herbivore in the absence of toxin, influences the effect of G. The larger $1/h$ is, the greater effect that a certain level G will have on population dynamics. Under biological considerations (e.g., $C(P) \geq 0$, $G \leq 1/h$, etc.), the following constraints will be imposed on G:

$$\frac{1}{4h} \leq G \leq \frac{1}{h}. \tag{4.5}$$

The fraction of encountered resources ingested, σ, is one of the parameters that will be adjusted. This parameter may also be considered as the herbivore's "consumption choice function" that depends on the plant density P. This is particularly important when multiple plant species (P_i, $i = 1, 2, \cdots, n$) are considered, in which case $\sigma_i(P_1, P_2, \cdots, P_n)$ represents the herbivore browsing preference to plant species i.

If σ is assumed to be constant, then when $1/(4h) < G < 1/(2h)$, $C(P)$ is unimodal, declining to an asymptote after reaching the peak value G at $P = P_m$:

$$C(P_m) = G, \quad \text{where } P_m = \frac{G}{e\sigma(1/2 - hG)} \tag{4.6}$$

(see the solid curve in Figure 4.1).

In the case of non-constant σ, the herbivores are able to avoid further consumption beyond the optimal rate of G, for any value of $P > P_m$. One example for this case is to define $C_1(P) = C(P, \sigma(P))$ with the function $\sigma(P)$ being given by

$$\sigma(P) = \begin{cases} \sigma_0 & \text{for } P \leq P_m, \\ \sigma_0 \dfrac{P_m}{P} & \text{for } P > P_m, \end{cases} \tag{4.7}$$

where $\sigma_0 > 0$ is a constant. In this case,

$$C_1(P) = \begin{cases} C(P, \sigma_0) & \text{for } P \leq P_m, \\ G & \text{for } P > P_m. \end{cases} \tag{4.8}$$

4.1.2　Dynamics for constant browsing preference

When the browsing preference is constant, i.e., $\sigma(P) = \sigma_0$, system (4.1) has two boundary equilibria at which $H = 0$:

$$E_0 = (P_0, H_0) = (0, 0), \quad E_K = (P_K, H_K) = (K, 0).$$

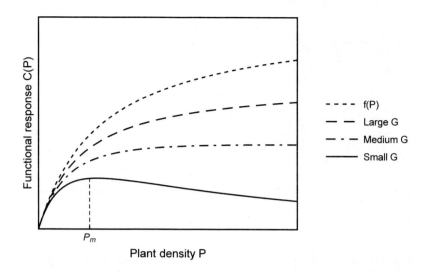

FIGURE 4.1: Graphs of the functional response $C(P)$ given in (4.2). For smaller G values (higher toxin level) $C(P)$ is unimodal and reaches its maximum at a finite value of P (the solid curve). For higher G values $C(P)$ is monotonically increasing with $f(P)$ as a limit.

To determine possible interior equilibrium points, it is easier to rewrite the equation for P in (4.1) as

$$\frac{dP}{dt} = C(P)\Big(g(P) - H\Big)$$

where

$$g(P) \doteq \frac{rP(1 - P/K)}{C(P)} = \frac{r(K - P)(1 + he\sigma_0 P)^2}{e\sigma_0 K[1 + e\sigma_0(h - 1/4G)P]}. \tag{4.9}$$

Then, system (4.1) is equivalent to the following system

$$\begin{aligned}
\frac{dP}{dt} &= C(P)\Big(g(P) - H\Big), \\
\frac{dH}{dt} &= BC(P)H - dH,
\end{aligned} \tag{4.10}$$

where $g(P)$ is given in (4.9). Denote a possible interior equilibrium of (4.1) by $E^* = (P^*, H^*)$ with $0 < P^* < K$ and $H^* > 0$. The properties of $g(P)$ (see Figure 4.2) and $C(P)$ will determine the number of interior equilibrium points.

4.1.2.1 Bifurcation analysis

Note that, at an interior equilibrium $E^* = (P^*, H^*)$, P^* satisfies the equation:

$$C(P^*) = d/B \quad \text{with } 0 < P^* < K \tag{4.11}$$

(see Figure 4.3). For each solution P^* of equation (4.11), H^* can be determined by

$$H^* = g(P^*) = \frac{r(K - P^*)(1 + he\sigma_0 P^*)^2}{e\sigma_0 K[1 + e\sigma_0(h - 1/4G)P^*]}. \tag{4.12}$$

Figure 4.3 illustrates that it is possible for equation (4.11) to have two solutions, $P_i^* (i = 1, 2)$ with $0 < P_1^* < P_2^* < K$. It is also clear that the number of solutions of equation (4.11) can be 0 (if $d/B > G$), 1 (if $d/B \leq G$ and $P_2^* \geq K$) or 2 (if $d/B < G$ and $P_2^* < K$).

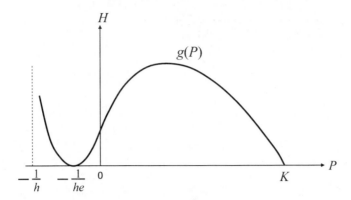

FIGURE 4.2: Plot of the function $g(P)$, which is the P-nullcline that determines interior equilibria. It has a single hump for $P \in (0, K)$.

As the main interest here is to explore the joint effect of the plant toxin and herbivore browsing on the outcomes of the plant-herbivore interaction, the bifurcation analysis is performed using two key parameters, G and $w = BG - d$. The parameter w represents the difference between the maximum possible energy intake by the herbivore (BG) and the energy loss due to mortality (d). Thus, w is a measure for the maximum individual fitness. The biologically reasonable region in the (G, w) plane is

$$\frac{1}{4h} < G < \frac{1}{2h}, \quad 0 < w < BG.$$

When both interior equilibria are possible, denote them by $E_1^* = (P_1^*, H_1^*)$ and $E_2^* = (P_2^*, H_2^*)$ with $0 < P_1^* < P_2^* < K$. The solutions P_i^* of equation (4.11) satisfy the following quadratic equation:

$$a_2 P^2 + a_1 P + a_0 = 0, \tag{4.13}$$

where

$$a_0 = -dG,$$

$$a_1 = e\sigma_0 G(B - 2hd),$$

$$a_2 = e^2 \sigma_0^2 (BhG - B/4 - dh^2 G).$$

Clearly, the number of biologically feasible equilibria depends on the values of model parameters. Consider the case when all parameters are fixed, except for G and w, which are varied. In Figure 4.3, various positions of the line d/B are shown intersecting with the functional response curve $C(P)$. As d decreases (or as w increases), the point P of intersection of d/B and $C(P)$ corresponding to the descending limb will ultimately exceed K, in which case the corresponding equilibrium E^* disappears and the number of interior equilibrium points changes from two to one. The dividing line between one and two interior equilibria defines a curve in the (G, w) plane, which is denoted by $w_K(G)$ and given by

$$w_K(G) = B\big(G - C(K)\big).$$

The curve $w = w_K(G)$ is depicted in the bifurcation diagram Figure 4.4, and it has the properties that (i) if $G < G_c$ (which is equivalent to $P_m < K$, see (4.14) for the expression of G_c), then both E_1^* and E_2^* exist for $w < w_K$ and only E_1^* exists for $w > w_K$; and (ii) if $G > G_c$ (or equivalently $P_m > K$), then only E_1^* exists for $w > w_K$ and there is no interior equilibrium for $w < w_K$.

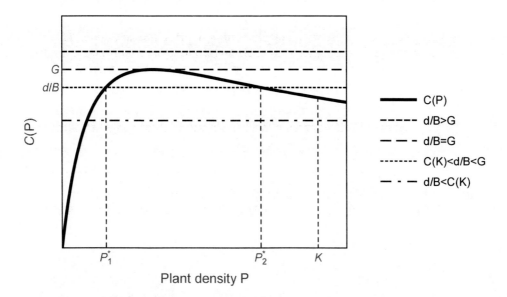

FIGURE 4.3: At an interior equilibrium $E^* = (P^*, H^*)$, P^* is an intersection of the line d/B with the curve $C(P)$ and $0 < P^* < K$. There is no intersection if $d/B > G$, one intersection if $d/B = G$ or $d/B < C(K)$, and two intersections (P_1^* and P_2^*) if $C(K) < d/B < G$.

In Figure 4.4, the line $w = 0$ determines whether there exists an interior equilibrium ($w > 0$) or not ($w < 0$). The curve $w_K(G)$ intersects with the line $w = 0$ at the critical value G_c, which is given by

$$G_c \equiv \frac{e\sigma_0 K}{2(1 + he\sigma_0 K)}. \tag{4.14}$$

The curve $w_K(G)$ divides the region into three sub-regions, denoted by I, II, and III, such that when (G, w) moves from one region to another, the number of interior equilibria changes. Results on the existence of interior equilibria in regions I, II, and III are summarized as follows:

I. In this region, the unique interior equilibrium is either $E_1^* = (P_1^*, H_1^*)$ (in this case $P_2^* > K$) or $E_2^* = (P_2^*, H_2^*)$ (in this case $P_1^* \leq 0$).

II. There is no interior equilibrium in this region.

III. In this region, there exist two interior equilibria: $E_i^* = (P_i^*, H_i^*)$ ($i = 1, 2$ with $0 < P_1^* < P_2^* < K$).

Stability of equilibria

Let $\bar{E} = (\bar{P}, \bar{H})$ denote an equilibrium of system (4.1), or the equivalent system (4.10). The Jacobian matrix at $\bar{E}$ is

$$J(\bar{E}) = \begin{pmatrix} C'(\bar{P})[g(\bar{P}) - \bar{H}] + C(\bar{P})g'(\bar{P}) & -C(\bar{P}) \\ BC'(\bar{P})\bar{H} & BC(\bar{P}) - d \end{pmatrix}. \tag{4.15}$$

It is easy to show that $E_0 = (0, 0)$ is always a saddle point. The Jacobian matrix at

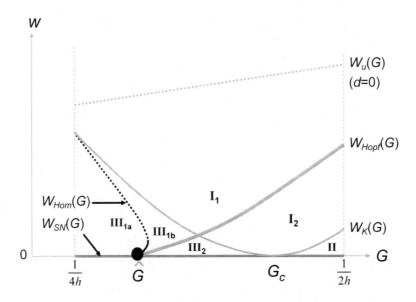

FIGURE 4.4: A bifurcation diagram in the (G, w) parameter plane for system (4.1) when $\sigma(P) = \sigma_0$ is a constant. It shows three bifurcation curves: $w_K(G)$, $w_{\text{Hopf}}(G)$ (Hopf bifurcation), and $w_{\text{Hom}}(G)$ (homoclinic bifurcation). These curves divide the feasible parameter region in the (G, w) plane into sub-regions: I_1, I_2, II, III_{1a}, III_{1b}, and III_2. Dynamics of the 2-D TDFRM (4.1) are different in these regions including existence and stability of equilibria and periodic solutions (see the text for more details).

$E_K = (K, 0)$ is

$$J(E_K) = \begin{pmatrix} C(K)g'(K) & -C(K) \\ 0 & BC(K) - d \end{pmatrix}. \tag{4.16}$$

From $C(K)g'(K) < 0$ (see Figure 4.2), a transcritical bifurcation occurs along the curve $w_K(G) = B[G - C(K)]$. E_K is an attracting node if $w < w_K(G)$, and it is a saddle if $w > w_K(G)$.

The Jacobian at $E_1^* = (P_1^*, H_1^*)$ is

$$J(E_1^*) = \begin{pmatrix} C(P_1^*)g'(P_1^*) & -C(P_1^*) \\ BC'(P_1^*)H_1^* & 0 \end{pmatrix}. \tag{4.17}$$

Note that the sign of $g'(P_1^*)$ may change, depending on whether P_1^* is on the left or right side of the point P_h, at which $g(P)$ achieves its maximum value. Clearly, $\det(J(E_1^*)) > 0$ as $C(P_1^*) > 0$ and $C'(P_1^*) > 0$. Notice that $\text{tr}(J(E_1^*)) = C(P_1^*)g'(P_1^*)$. Hence, E_1^* is a stable focus if $g'(P_1^*) < 0$ ($P_1^* > P_h$) (Figure 4.5a) and an unstable focus if $g'(P_1^*) > 0$ ($P_1^* < P_h$) (Figure 4.5b). Thus, the stability of E_1^* switches at the point where $P_1^* = P_h$, at which $g'(P_1^*) = 0$. This determines an equation for G and w. Note that $BC(P_1^*) - d = 0$ also determines an equation for G and w. Thus, a Hopf bifurcation may occur if P_1^* satisfies both of the equations:

$$g'(P_1^*) = 0 \quad \text{and} \quad BC(P_1^*) - d = 0,$$

which determine the Hopf bifurcation curve, denoted by $w = w_{\text{Hopf}}(G)$. The interior equilibrium E_1^* is stable for $w < w_{\text{Hopf}}$ and unstable for $w > w_{\text{Hopf}}$ (see Figure 4.4). A more detailed proof of this result can be found in [226].

It can be shown that the curve $w_{\mathrm{Hopf}}(G)$ intersects with $w = 0$ at the critical point $\hat{G}$, which is given by

$$\hat{G} = \frac{e\sigma_0 K}{2(2 + he\sigma_0 K)} < G_c.$$

The Hopf curve increases with G, lies below the line $w_u(G)$ for all $\hat{G} \le G \le 1/(2h)$, and intersects with the curve $w_K(G)$ at only one point (see Figure 4.4).

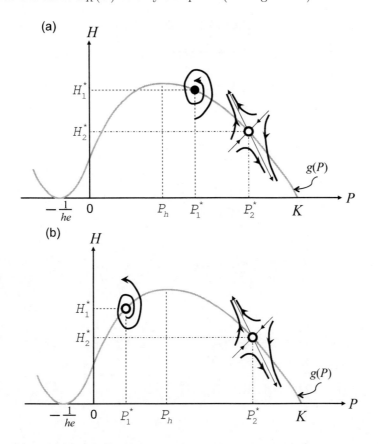

FIGURE 4.5: Stabilities of the two interior equilibria, $E_i^* = (P_i^*, H_i^*)$, $i = 1, 2$, are related to the properties of the function $g(P)$. E_1^* is a stable focus if $P_1^* > P_h$ (or $g'(P_1^*) < 0$, see (a)), and it is an unstable focus if $P_1^* < P_h$ (or $g'(P_1^*) > 0$, see (b)).

For the stability of E_2^*, noticing that $P_2^* > P_m$ and hence $C'(P_2^*) < 0$ (see Figure 4.3), it follows that

$$\det(J(E_2^*) = BC'(P_2^*)C(P_2^*)H_2^* < 0.$$

Therefore, E_2^* is a saddle point whenever it exists.

The curve $w_{\mathrm{Hopf}}(G)$ divides region I into regions I_1 and I_2 and divides region III into regions III_1 ($=\mathrm{III}_{1a} \cup \mathrm{III}_{1b}$) and III_2. The point $(\hat{G}, 0)$, where the curve $w_{\mathrm{Hopf}}(G)$ intersects with $w = 0$, is a cusp point of co-dimension 2, implying the possibility of a homoclinic bifurcation (more details can be found in [226]). The homoclinic curve $w_{\mathrm{Hom}}(G)$ is indicated in Figure 4.4, which further divides region III_1 into two sub-regions, III_{1a} and III_{1b}.

Figure 4.6 depicts the stability of possible equilibria of system (4.1) in each of the regions shown in Figure 4.4. For example, In region I_1, there is a unique interior equilibrium E_1^*, which is an unstable focus, and solutions converge to a stable periodic solution as time tends

TABLE 4.1: Local stability results when σ is constant.

	I_1	I_2	II	III_{1a}	III_{1b}	III_2
E_0	Saddle	Saddle	Saddle	Saddle	Saddle	Saddle
E_1^*	Unstable	**Stable**	DNE	Unstable	Unstable	**Stable**
E_2^*	DNE	DNE	DNE	Unstable	Saddle	Saddle
E_K	Saddle	Saddle	**Stable**	**Stable**	**Stable**	**Stable**
PS	**Stable**	DNE	DNE	DNE	**Stable**	DNE

DNE: Does not exist; PS: Periodic solution

to infinity (see Figure 4.6a). In region I_2, the unique interior equilibrium is a stable focus (Figure 4.6b). Figures 4.6d–f show the cases in region III, in which both interior equilibria exist with E_2^* being a saddle and E_1^* being either an unstable focus (regions III_{1a} and III_{1b}) or a stable focus (region III_2). A stable periodic solution is present in region III_{1b}. Notice that for parameters in regions III_{1b} and III_2, bistability is possible (two local attractors). Results for the stability of equilibria in these regions are also summarized in Table 4.1.

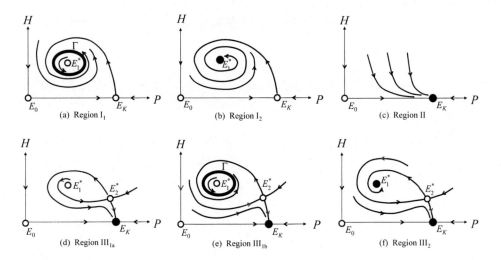

FIGURE 4.6: Phase portraits in each of the regions shown in Figure 4.4. The circle and solid dot represent unstable and stable equilibria, respectively.

Biological relevance of the bifurcations

The bifurcation analysis provides insights into how the plant toxicity may influence the plant-herbivore interactions. Figure 4.7 shows results of numerical simulations for several scenarios based on the bifurcation diagram Figure 4.4. Five sets of (G, w) values are selected as shown in Figure 4.7a.

If both G and w are decreased (but d is fixed at a constant value d_0) along a diagonal straight line from the middle right-hand corner of the bifurcation diagram (Figure 4.7a) toward the lower left-hand corner, the effects of G alone on the system behavior can be observed. The system undergoes a series of changes leading from dominance of simple Holling type 2 dynamics to dominance by the effect of the toxicant. For the larger value G with (G, w) in region I_2, the single interior equilibrium E_1^* is stable (see Figure 4.7b). As G and w decrease, moving from region I_2 to I_1, the single stable interior equilibrium E_1^* becomes unstable, producing a limit cycle (marked by the "sps" in Figure 4.7c), a result

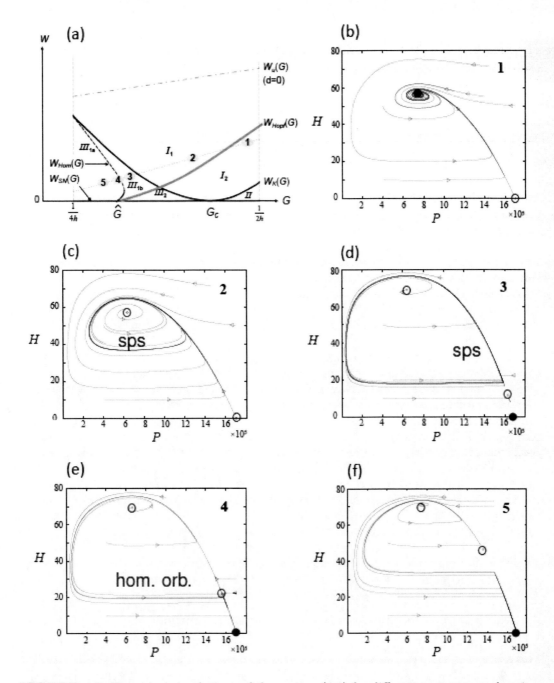

FIGURE 4.7: Numerical simulations of the system (4.1) for different parameter values in the (G, w) plane, labeled as 1-5 in (a). In the phase portraits (b)-(f), a solid dot denotes a stable equilibrium and a circle denotes an unstable equilibrium. A stable periodic solution is denoted by "sps," and a homoclinic orbit is marked by "hom. orb."

of increasing toxicant effect. As G and w continue to decrease and move into region III_{1b}, the saddle point moves into the positive quadrant, with $P_2^* < K$, so there are now two interior equilibria (Figure 4.7d). Biologically, the behavior for decreasing G means that in this region of parameter space the individual herbivores may have trouble dealing with the amount of toxicant that they are ingesting, and the possibility exists of the herbivore being in an alternative state of poor physiological condition; the saddle point E_2^*. The limit cycle around the point E_1^* can be changed to a stable node by holding G fixed and decreasing w into region III_2. In region III the consumer population is in some danger of going to extinction due to the toxicant ingested from the plant, because the herbivore has no control over its rate of ingestion, which is completely determined by the density of plant food items, P. The fate of the herbivore population depends on the initial values of P and H. If P starts at a large value, then, because the herbivore feeding rate increases with plant density, the herbivores will consume themselves to extinction and the system will approach the boundary equilibrium $P = K$, $H = 0$. Also, even if P is initially small, but H also starts out very small, the herbivore may ultimately go to extinction. This is because an initially small value of H allows P to grow long enough to reach density levels that lead to the herbivore population dying out from toxicity. A relatively small initial value of P, together with relatively large H, however, could allow the system to be captured by the stable limit cycle around E_1^*.

Further decreases in G and w from region III_{1b} take the system across the homoclinic threshold (Figure 4.7e) into region III_{1a}. In this region, the limit cycle is close enough to the saddle that it is no longer possible for a stable limit cycle to exist, as it is "captured" by the saddle point. Now trajectories from every point in the (P, H)-plane lead toward the saddle point and ultimately to the boundary point $E = (K, 0)$, where the herbivore is extinct (Figure 4.7f). Ecologically this means that the level of toxicant in the plant is so high that the herbivore cannot survive on the plant. It can obtain enough energy to survive from the plant only at the cost of absorbing so much toxicant that it dies.

The above results suggest that, in the case when there are two interior equilibria, one state is the healthy state in which $P = P_1^*$ is low enough that the herbivore's food intake is below its peak level. In this state herbivore grazing exerts a top-down effect, controlling the plant biomass down to this level. The other equilibrium, at a higher plant biomass level, P_2^*, represents a weakened state of the herbivore with a smaller population value. In this state the herbivore density is depressed, as it is controlled by the plant defenses. Its ingestion rate is decreased because of physiological damage from the toxin. The herbivore population is not trapped in that state, because E_2^* is a saddle point (unstable). A slight decrease in P would produce positive feedback mechanisms taking the herbivore back to E_1^*, either a stable equilibrium or stable limit cycle due to the bistability. But a slight increase in P would push the herbivore population in the opposite direction, toward extinction.

4.1.3   Dynamics for non-constant browsing preference

Although it is not difficult to imagine that animals, when encountering food at high levels, may eat to a point that is detrimental to them, as noted above it is also likely that many species are adapted for controlling their rates of ingestion. In this section, we consider the case where the herbivore's "consumption choice function", $\sigma(P)$, is given in equation (4.7) and the corresponding functional response $C_1(P)$ is given in equation (4.8). The bifurcation analysis uses G and w as bifurcation parameters.

System (4.1) with $C_1(P)$ also has two boundary equilibria at which $H = 0$: $E_0 = (P_0, H_0) = (0, 0)$ and $E_K = (P_K, H_K) = (K, 0)$. However, there is only one possible interior equilibrium, which we denote by $E^* = (P^*, H^*)$ with $0 < P^* < K$, and $H^* > 0$. The conditions for the existence and stability of all these equilibria depend on the magnitudes of G and w when other parameters are fixed. The analysis of existence and stability of equilibrium points generates two bifurcation curves in the (G, w) parameter plane, which are labeled in Figure 4.8 as $w = w_K(G)$ and $w = w_{\mathrm{Hopf}}(G)$. These curves have similar meanings as those in Figure 4.4. For example, the curve $w = w_K(G)$ separates the regions based on the number of equilibrium points and their stability. The curve $w_{\mathrm{Hopf}}(G)$ indicates a Hopf bifurcation such that periodic solutions may exist for (G, w) above the curve. More detailed descriptions are stated in the following results, which are also summarized in Figure 4.8 and Table 4.2.

(i) If $G > G_c$ and $w < w_K$ (region II in Figure 4.8), then there is no interior equilibrium and the boundary equilibrium E_K is locally asymptotically stable.

(ii) If $G > \hat{G}$ and $w_K < w < w_{\mathrm{Hopf}}$ (region I_2), then E_K is unstable and a unique interior equilibrium E^* exists, which is locally asymptotically stable.

(iii) If $w > 0$ and $w > w_{\mathrm{Hopf}}$ (regions I_1 & III), then E_K and the unique interior equilibrium E^* are both unstable, and a stable periodic solution exists via Hopf bifurcation.

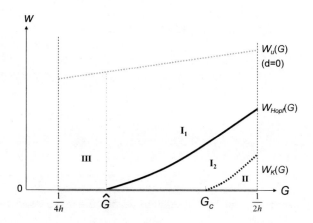

FIGURE 4.8: Bifurcation diagram for system (4.1) similar to Figure 4.4 but for the case of non-constant $\sigma(P)$. The bifurcation curves $w_{\mathrm{Hopf}}(G)$, $w_K(G)$, and $G = \hat{G}$ divide the parameter region into sub-regions: I_1, I_2, II, and III. It shows that the existence and stability of equilibria and periodic solution are different from the case when σ is constant (see the text for more details).

TABLE 4.2: Local stability results when $\sigma(P)$ is not constant.

	I_2	II	I_1 & III
E_0	Saddle	Saddle	Saddle
E^*	**Stable**	DNE	Unstable
E_K	Saddle	**Stable**	Saddle
PS	DNE	DNE	**Stable**

DNE: Does not exist; PS: Periodic solution.

4.1.4 Global dynamics of the 2-dimensional TDFRM

For the study of spatial extensions of the model (4.1) (see Section 4.4), it is necessary to identify conditions for the global dynamics of the system (4.1) in the absence of periodic solutions. This is investigated in [53]. The analysis focuses on the case of constant σ and assumes the existence of an interior equilibrium, which requires the condition

$$\max_{0 \leq P \leq K} C(P) > \frac{d}{B}. \tag{4.18}$$

Let $g(P)$ be the same function as defined in (4.9). The number of positive equilibria of system (4.1) is determined by the number of solutions P of the equation

$$BC(P) = d \tag{4.19}$$

in the interval $(0, K)$. If the system has a unique interior equilibrium, denote it by $E_1^* = (P_1^*, g(P_1^*))$. If there are two interior equilibria, denote them by $E_i^* = (P_i^*, g(P_i^*))$, $i = 1, 2$, with $P_1^* < P_2^*$, as shown in Figure 4.5. Note that E_1^* is either locally asymptotically stable (if $g'(P_1^*) < 0$) or unstable (if $g'(P_1^*) > 0$), while E_2^* is always a saddle whenever it exists. Consider the case when E_1^* is stable, i.e., the following condition holds

$$g'(P_1^*) < 0. \tag{4.20}$$

Before stating the result of global stability of E_1^*, the following result is helpful for excluding closed orbits. For easy of presentation, consider the following system equivalent to (4.1):

$$P' = C(P)\Big[g(P) - H\Big],$$
$$H' = dC(P)M(P)H, \tag{4.21}$$

where $M(P) = B/d - 1/C(P)$.

Theorem 1 *Under the condition (4.20), system (4.21) has no closed orbit in the region* $\mathbf{R}_+^2 = \{(P, H) : P \geq 0, H \geq 0\}$.

Proof: Suppose that (4.21) has a closed orbit $\gamma = \{(P(t), H(t)) : t \in [0, \omega]\}$ in $\mathbf{R}_+^2$. Then, by analyzing the vector field of (4.21) (see Figure 4.9) it can be shown that $P(t) < \min\{P_2^*, K\}$ for $t \in [0, \omega]$. In addition, without loss of generality, assume that there is an $\omega_1 \in (0, \omega)$ such that

$$P(0) = P(\omega_1) = P(\omega) = P_1^*,$$
$$P(t) < P_1^*, \quad t \in (0, \omega_1),$$
$$P(t) > P_1^*, \quad t \in (\omega_1, \omega).$$

Let $\xi(t) = G(P(t))$. Then $P(t) = \theta(\xi(t))$ and

$$\xi'(t) = C(\theta(\xi(t)))\big\|M(\theta(t))\big\|[g(\theta(\xi(t))) - H(t)],$$
$$H'(t) = dC(\theta(\xi(t))M(\theta(t))H(t). \tag{4.22}$$

From (4.22),

$$\frac{d\xi}{dH} = -\frac{[g(\theta(\xi(t))) - H(t)]}{BH(t)}, \quad t \in [0, \omega_1],$$
$$\frac{d\xi}{dH} = \frac{[g(\theta(\xi(t))) - H(t)]}{BH(t)}, \quad t \in [\omega_1, \omega]. \tag{4.23}$$

Noticing that

$$BC(P) - d < 0, \quad P \in (0, P_1^*) \quad \text{and}$$

$$BC(P) - d > 0, \quad P \in (P_1^*, \min\{P_2^*, K\}),$$

and using the definition of $M(P)$ and the expression (4.22), it can be obtained that

$$
\begin{aligned}
M(P(t)) &< 0, \quad t \in (0, \omega_1) \quad \text{and} \\
M(P(t)) &> 0, \quad t \in (\omega_1, \omega).
\end{aligned}
\tag{4.24}
$$

It follows from (4.22) and (4.24) that $H(t)$ is strictly decreasing in the interval $[0, \omega_1]$ and strictly increasing in $[\omega_1, \omega_2]$. Let $H_0 = H(0) = H(\omega)$ and $H_1 = H(\omega_1)$, then the inverse functions can $t_1^{-1} : [H_0, H_1] \to [0, \omega_1]$ and $t_2^{-1} : [H_0, H_1] \to [\omega_1, \omega]$ can be defined as follows:

$$
\begin{aligned}
t &= t_1^{-1}(u) \in [0, \omega_1] \quad \text{if and only if} \quad u = H(t), \\
t &= t_2^{-1}(v) \in [\omega_1, \omega] \quad \text{if and only if} \quad v = H(t).
\end{aligned}
\tag{4.25}
$$

Let $\xi_1(H) = -\xi(t_1^{-1}(H))$ and $\xi_2(H) = \xi(t_2^{-1}(H))$, then

$$
\begin{aligned}
\xi_1(H_0) &= \xi_2(H_0) = \xi_1(H_1) = \xi_2(H_1) = 0, \\
0 &< \xi_i(H) \quad \text{for} \quad H_0 < H < H_1.
\end{aligned}
\tag{4.26}
$$

Moreover, the chain rule in $H_0 < H < H_1$ can be used to compute the derivative of the inverse function. Using (4.22) and (4.23) it can be obtained that

$$
\begin{aligned}
\frac{d\xi_1}{dH} &= \frac{g(\theta(-\xi_1)) - H}{BH}, \\
\frac{d\xi_2}{dH} &= \frac{g(\theta(\xi_2)) - H}{BH}.
\end{aligned}
\tag{4.27}
$$

On the other hand, from Proposition 3.1 in [53] it follows that $g(\theta(-\xi)) > g(\theta(\xi))$ for all $0 < \xi < \xi_m$. Hence, the inequalities in (4.27), in conjunction with the use of comparison principle, yield that $\xi_1(H) > \xi_2(H)$ for all $H \in (H_0, H_1]$. In particular, $\xi_1(H_1) > \xi_2(H_1)$, which contradicts (4.26). This completes the proof.

The main result for the global dynamics of (4.1) is stated in the following theorem.

Theorem 2 *Suppose that condition (4.20) is satisfied.*

1. *If E_1^* is a unique positive equilibrium of (4.1) then all positive solutions of the system converge to E_1^* as $t \to \infty$.*

2. *If both E_1^* and E_2^* exist, then E_2^* is a saddle point. Moreover, the stable manifold of E_2^* divides the first quadrant of the P-H plane into two sub-regions, Ω_1 and Ω_K (see Figure 4.10). Every solution through a point in Ω_1 converges to E_1^*, while every solution with the initial point in Ω_K converges to the boundary equilibrium $E_K = (K, 0)$ as $t \to \infty$.*

Theorem 2 can be proved by applying Theorem 1 and the Poincaré–Bendixon theorem. The proof can be found in [53]. The global stability result stated in this theorem can be very helpful for identifying situations in which the herbivore population may tend to extinction (e.g., when the initial condition lies in the shaded region Ω_K; see Figure 4.10). Note that

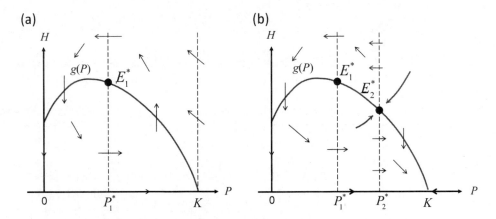

FIGURE 4.9: (a) Vector field with a single interior equilibrium E_1^*. (b) Vector field with two interior equilibria E_1^* and E_2^*.

the result stated in the theorem holds only for parameters in the region in which the system (4.1) does not have a limit cycle. In the case when the system has a stable period solution, a similar result may be possible but it has not yet been proved.

Theorem 2 shows that the sign of $g'(P_1^*)$ provides key information for the global dynamics of system (4.1). If $g'(P_*) < 0$ then the system is dissipative, i.e., each positive solution will converge to an equilibrium point. In particular, if there is a unique interior equilibrium point E_1^*, then the local stability of E_1^* implies global stability.

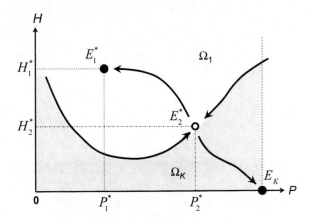

FIGURE 4.10: Global attraction regions of system (4.1) described in Theorem 2. The stable manifold of $E_2^* = (P_2^*, H_2^*)$ separates the two regions of attraction, Ω_1 (unshaded) and Ω_K (shaded), for the stable equilibria $E_1^* = (P_1^*, H_1^*)$ and $E_K = (K, 0)$.

The situation would be more complicated under the assumption that $g'(P_*) > 0$, because in this case the system of interest would no longer be dissipative but oscillatory. Previous results suggest that if the interior equilibrium point E_1^* is unique and unstable, then there is a unique positive periodic solution which attracts all positive solutions except the equilibrium point E_1^*. This is a highly challenging problem, and the mathematical results and approaches are likely to have useful applications in the study of plant-herbivore interactions.

More results on the global dynamics of system 4.1 are presented in [389]. The bifurcation results presented in Section 4.1.2 are primarily local including the occurrence of Hopf bifurcation, and the homoclinic bifurcation is based mainly on numerical simulations. These results are extended in [389], in which a global bifurcation analysis is conducted and it is shown that the stable limit cycles from the Hopf bifurcation exist globally and terminate with the appearance of a homoclinic orbit.

4.2 A 3-dimensional TDFRM with two plant species

The results presented in Section 4.1 show that the TDFRM can reasonably describe a variety of possible dynamics of an herbivore population feeding on a single plant species. However, it is more typical for mammalian herbivores to feed on a variety of plants [112, 381, 382]. The 2-D TDFRM (4.1) includes only a single plant species. In this section, we investigate the effect of toxicants when the herbivore has a choice of two plant species to feed on. In this case, σ is generalized to σ_i ($i = 1, 2$) to describe the relative effort exerted on feeding on each particular plant species i. Two situations will be considered. One situation is when the toxins in the plants are distinct enough to be non-interactive, so that the amount of one toxin that an herbivore can tolerate is not affected by the presence of the other toxin. The other extreme situation is when the negative effects of the two toxins are perfectly additive in their negative effects. Intermediate levels of interaction can occur as well, but will not be modeled here. Nutrient-toxin interactions will also be ignored.

The first extension of the 2-D model (4.1) is to include two plant species to model how interspecific plant competition, herbivory, and a plant's toxic defenses against herbivores may affect vegetation dynamics. Herbivores can affect plant populations and plant communities by altering plant competition, facilitating the invasion of extant plant communities by new plants (invaders), and modifying the trajectory (rate and outcome) of plant community succession [75, 140]. In response to herbivory, plants have evolved a variety of defenses, many of which are in the form of chemicals that are toxic to herbivores (reviewed by Stamp [83, 341]). In counter-response, many herbivores have evolved offensive tactics as a means of counteracting plant defenses (reviewed by Karban and Agrawal [183]). General models that incorporate the interplay of plant defense and herbivore offense are needed if ecologists are to better understand the implications of these complex plant-herbivore interactions for community dynamics. We will also consider how herbivory and plant competition interact to influence community dynamics when the herbivore adopts a foraging strategy consisting of a variable effort that changes through time to maximize fitness.

The objective here is to first examine how the dynamics of two competing plant species is affected when herbivory is added but there are no anti-herbivore toxins, and next how the dynamics is affected when different concentrations of either non-interacting or additive toxins are included in the two different plant species. There exists a body of empirical research that shows that consumers often feed in a positive frequency-dependent manner, such that the more abundant prey species is disproportionately represented in the diet (e.g., [121]). Such positive frequency-dependent feeding may result naturally from the forming of a search image for the more abundant prey type, which may be adaptive [158, 159]. Theoretical studies indicate that such behavior by the consumer can contribute to species coexistence (e.g., [71]). Thus, from such studies it is expected that when herbivores forage adaptively on plants, they should tend to promote coexistence through a certain amount of switching to more abundant prey, thus preventing the extinction of inferior plant competitors. However, these studies did not take into account the effect of plant toxins. In [98], the effects that plant toxins have on this traditional view of the role of herbivores on plant communities is examined. The study is organized around following questions and hypotheses:

1. How do plant toxins change these dynamics when herbivores forage adaptively? The first hypothesis (Hypothesis 1) is that when toxins are present in different concentrations in the competing plants, the effects of toxins can cause a reversal in the behavior of the herbivore, toward concentrating foraging effort on the rarer plant species (thus acting in a depensatory manner), when either the toxins are non-interacting or the rarer species is less toxic.

2. In successional processes in the presence of herbivores, the second hypothesis (Hypothesis 2) is that less defended fast-growing species will generally be replaced over time by more defended slower-growing species.

3. The third hypothesis (Hypothesis 3) is that the presence of a resident species that is highly toxin-defended will generally inhibit the invasion of a new species unless the invading species is itself highly toxic.

The model formulation, analyses, and validation are presented in the following sections.

4.2.1 The model with two plant species

Let P_1, P_2, and H denote the biomass densities of edible components of plant species 1 and 2, and of the herbivore, respectively. Let G_i, h_i, r_i, e_i, B_i, d, and K_i ($i = 1, 2$) denote the corresponding parameters in the case of a single plant species with the subscript representing species i, and c_{ij} denote the coefficients of competitive effect of plant species j on i ($i \neq j$). The equations for the 3-dimensional TDFRM (which will be referred to as 3-D TDFRM) are:

$$\frac{dP_1}{dt} = r_1 P_1 \left(1 - \frac{P_1 + c_{12} P_2}{K_1} \right)$$

$$- \frac{e_1 \sigma_1 P_1 H}{1 + h_1 e_1 \sigma_1 P_1 + h_2 e_2 \sigma_2 P_2} \left(1 - \frac{e_1 \sigma_1 P_1}{4 G_1 (1 + h_1 e_1 \sigma_1 P_1 + h_2 e_2 \sigma_2 P_2)} \right),$$

$$\frac{dP_2}{dt} = r_2 P_2 \left(1 - \frac{P_2 + c_{21} P_1}{K_2} \right)$$

$$- \frac{e_2 \sigma_2 P_2 H}{1 + h_1 e_1 \sigma_1 P_1 + h_2 e_2 \sigma_2 P_2} \left(1 - \frac{e_2 \sigma_2 P_2}{4 G_2 (1 + h_1 e_1 \sigma_1 P_1 + h_2 e_2 \sigma_2 P_2)} \right),$$

$$\frac{dH}{dt} = \frac{B_1 e_1 \sigma_1 P_1 H}{1 + h_1 e_1 \sigma_1 P_1 + h_2 e_2 \sigma_2 P_2} \left(1 - \frac{e_1 \sigma_1 P_1}{4 G_1 (1 + h_1 e_1 \sigma_1 P_1 + h_2 e_2 \sigma_2 P_2)} \right)$$

$$+ \frac{B_2 e_2 \sigma_2 P_2 H}{1 + h_1 e_1 \sigma_1 P_1 + h_2 e_2 \sigma_2 P_2} \left(1 - \frac{e_2 \sigma_2 P_2}{4 G_2 (1 + h_1 e_1 \sigma_1 P_1 + h_2 e_2 \sigma_2 P_2)} \right) - dH,$$

which, in short notation, can be written as

$$\frac{dP_i}{dt} = r_i P_i \left(1 - \frac{P_i + c_{ij}P_j}{K_i}\right) - C_i(P_1, P_2)H, \quad i, j = 1, 2, \ i \neq j,$$

$$\frac{dH}{dt} = \sum_{i=1}^{2} B_i C_i(P_1, P_2)H - dH,$$

$$(4.28)$$

where $C_i(P_1, P_2)$ are defined as

$$C_i(P_1, P_2) = f_i(P_1, P_2)\left(1 - \frac{f_i(P_1, P_2)}{4G_i}\right), \quad i = 1, 2, \tag{4.29}$$

$$f_i(P_1, P_2) = \frac{e_i \sigma_i P_i}{1 + h_1 e_1 \sigma_1 P_1 + h_2 e_2 \sigma_2 P_2}, \quad i = 1, 2. \tag{4.30}$$

The toxin-determined functional response functions $C_i(P_1, P_2)$, $i = 1, 2$, in (4.29) start with the Holling type 2 functional response (first factor in equation (4.29)) for foraging on two species, which is denoted by $f_i(P_1, P_2)$, and is extended to include the effects of each of the two plant species containing its own toxin. The second factor in equation (4.29), $1 - f_i(P_1, P_2)/(4G_i)$, accounts for the negative effect of toxin. The plant species i with a small value of G_i has a strong effect of toxicity on herbivore foraging.

4.2.2   Validation of the model

To validate the TDFRM (4.28), it is applied to data from the Alaska Bonanza Creek Long-Term Ecological Research Project (BNZ LTER). Feeding experiments and herbivore exclusion experiments are relevant to the hypothesis that herbivory can lead succession toward more heavily defended and often slower-growing species. Such experimental studies in the boreal forest have been performed in the winter on mammalian herbivores such as snowshoe hare (*Lepus americanus*) and moose (*Alces alces*) to determine whether the per capita daily intake of browse (woody twigs, bark) is related to successional state (feeding experiments–[38]) and if selective browsing can alter the rate and trajectory of forest succession (exclusion experiments–[51, 192, 287]). Similar information has been provided by the work of Clay in temperate grasslands [65, 66, 67].

In 1987 exclosures-control pairs plots (20m×30m plots separated by 20m buffer strip; replicated 7 times) were randomly located in the mid-colonization stage of forest succession [62] (see Figure 4.11). The river terrace the plots were established on was about 12 years old. The plots were subdivided into 5 strata (4m wide) oriented parallel to the river and a permanent 2m^2 rectangular quadrat was randomly located within each strata so subsequent vegetation surveys were done at the same spot. The quadrats were used to obtain the ratios of willow/alder used to test the TDFRM and the Holling type 2 functional response predictions. Because there was very little litter in the first few years after sedimentation began twig counts were used to obtain the ratios in "model years 14, 17, and 20." Because of the time required to make census twigs, leaf litter was used in years 20, 26, and 32. In year 20 then estimates of the alder/willow ratios obtained by the twig census method and the leaf litter method were similar indicating use of the two methods does not affect the conclusions.

The results of the experiment are illustrated in Figure 4.11. Figure 4.11a demonstrates the mid-colonization stage of the BNZLTER exclosures and browsed control plots in 1985 ([192]; sensu Chapin et al. [62]). Figure 4.11b and c show the plant densities after 8 and 15 years, respectively. The ratios of alder to willow in both scenarios are illustrated in Figure 4.11d.

FIGURE 4.11a: A field experiment of Tanana floodplain showing the seedling stage of alder and willow in 1985 (time $t = 0$).

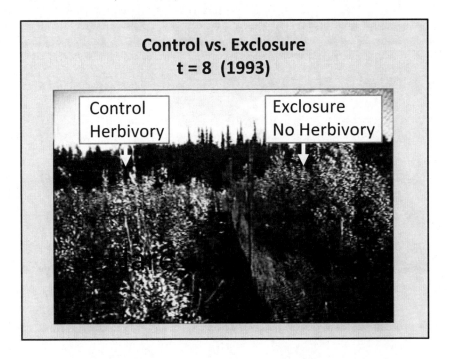

FIGURE 4.11b: A field experiment of Tanana floodplain showing changes on densities of alder and willow with and without browsing 8 years later ($t = 8$).

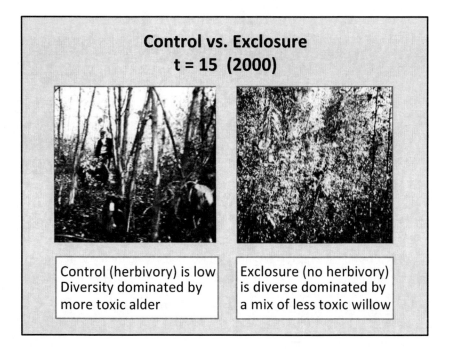

FIGURE 4.11c: A field experiment of Tanana floodplain showing changes on densities of alder and willow with and without browsing 15 years later ($t = 15$).

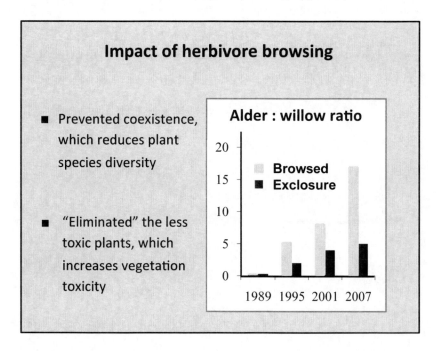

FIGURE 4.11d: Outcomes of the field experiment of Tanana floodplain. The insert plot shows the ratio of alder to willow at the end of the experiment with and without browsing.

Figure 4.12 presents the outcomes of the two models, one is the Holling type 2 functional response model (HT2FRM) and the other one with TDFRM. In Figure 4.12, the TDFRM is parameterized for the Tanana Floodplain system and computed the expected ratio of alder/willow from 1987-2007 for both the browsed control plots and the unbrowsed exclosure plots. The Holling type 2 model is also used to simulate browsing's effect on the ratio. The TDFRM (thick solid line) very accurately simulated changes in community composition, whereas the HT2FRM could not be fit reasonably to the data. In the presence of herbivores, the more toxic alder continued increasing in abundance relative to the less toxic willow. In the absence of herbivores (exclosure), the two plant species approached an equilibrium at a lower alder:willow ratio. In other words, herbivores speeded the elimination of the less toxic early successional willow. In this figure, several parameters are first estimated including the plant growth r_i, carrying capacity K_i, and the competition coefficients c_{ij} $(i, j = 1, 2, i \neq j)$ by fitting the model to the plant ratio data without herbivore browsing (in which case the two models are identical). The TDFRM and HT2FRM are then compared with herbivore browsing (the difference between the two models is that one model includes plant toxins explicitly while the other does not) by looking at the plant ratios in comparison with the plant ratio data with browsing. The parameter values used in the simulations are $r_1 = 0.0016, r_2 = 0.0017, K_1 = 50000, K_2 = 140000, c_{12} = 0.17, c_{21} = 0.2, e_1 = e_2 = 0.0001, h_1 = 0.01, h_2 = 0.06, G_1 = 98, G_2 = 5, B_1 = 0.00034, B_2 = 0.0003, d = 1/(3.5 \times 365)$.

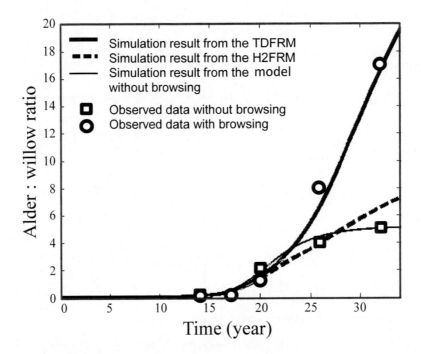

FIGURE 4.12: Comparison of simulations between TDFRM and HT2FRM based on the LTER data. Simulations of the two models are fit to the measured ratio alder/willow with and without browsing as shown in Figure 4.11d. See the main text for more explanations.

4.2.3   Analytic results for fixed allocation of foraging effort σ_i

The 3-dimensional TDFRM (4.28) can also be used to address the question of whether herbivores in the absence of plant toxins prevent the extinction of an inferior competitor. In the case of two plant species, the parameters σ_1 and σ_2 can be interpreted as the fractions of effort or time applied to foraging for the respective types of plant species. This differs from the "one plant species" case, where σ represented the fraction of encountered items of prey biomass that are ingested. Two cases will be considered here: one is when plant toxins are absent and the other is when toxins are included in the model (4.28).

4.2.3.1   Analysis of the model in the absence of toxins

Assume that the herbivores do not feed adaptively but have a strategy of fixed σ_1 and σ_2. (These results are intended to illustrate the alternative assumption to that when the herbivores forage adaptively.) As for the case of no herbivore, it is assumed that $c_{12} < 1 < c_{21}$ (under which species 1 will always exclude species 2 if $K_1 = K_2$). Here, the analysis is conducted for a more general case that allows for different K_i $(i = 1, 2)$.

To explore the conditions under which species 2 can invade in an environment where species 1 is already established, the nontrivial equilibrium $E_H^* = (P_{1H}^*, 0, H_H^*)$ (the subscript H for Holling) can be considered, where

$$P_{1H}^* = \frac{d}{e_1\sigma_1(B_1 - h_1 d)}, \quad H_H^* = \frac{r_1(1 - P_{1H}^*/K_1)}{e_1\sigma_1/(1 + h_1 e_1 \sigma_1 P_{1H}^*)}.$$

Clearly, $P_{1H}^* > 0$ when $d < B_1/h_1$. Thus, E_H^* is biologically feasible if $P_{1H}^* < K_1$. This requires that σ_1 is bounded below by some positive number which is denoted by σ_{1a}. The Jacobian matrix at E_H^* has two eigenvalues (either real or complex) with a negative real part. The third eigenvalue is negative if and only if

$$\frac{r_1}{r_2} > \frac{e_1\sigma_1(1 - c_{21}P_{1H}^*/K_2)}{e_2\sigma_2(1 - P_{1H}^*/K_1)}. \tag{4.31}$$

Therefore, the equilibrium E_H^* is locally asymptotically stable if the condition (4.31) holds. If the inequality (4.31) is reversed then E_H^* becomes unstable, in which case the invasion of species 2 is expected.

Notice that $P_{1H}^* = P_{1H}^*(\sigma_1)$ is a function of σ_1 and that $\sigma_2 = 1 - \sigma_1$. Thus, the right-hand side of the inequality (4.31) defines a function of σ_1, which is denoted by $F_1(\sigma_1)$. The curve corresponding to $F_1(\sigma_1)$ is shown in Figure 4.13, which intersects the σ_1 axis at σ_1^*. The threshold condition (4.31) implies that in the region above the curve $F_1(\sigma_1)$ invasion of species 2 is impossible if it starts at a low density.

By using a symmetric argument (switching between P_1 and P_2) another function $F_2(\sigma_1)$ can be identified, which determines the stability for the symmetric equilibrium at which species 1 is absent. It requires σ_1 to be bounded above by some positive number less than 1 which is denoted by σ_{1b}. The curve of $F_2(\sigma_1)$ is shown in Figure 4.13. Thus, species 2 can exclude species 1 for $(\sigma_1, r_1/r_2)$ below both curves. In the region below $F_1(\sigma_1)$ and above $F_2(\sigma_1)$, coexistence of the two plant species is expected.

4.2.3.2   Analysis of the model with toxins included

Now consider the case when plant toxicity is included in the model. (Again, these results are intended to illustrate the alternative assumption to that considered in the previous sections, where the herbivores foraged adaptively.) Here, the analysis of invasion criteria is presented for system (4.28). Consider the non-trivial equilibrium $E_T^* = (P_{1T}^*, 0, H_T^*)$ (the

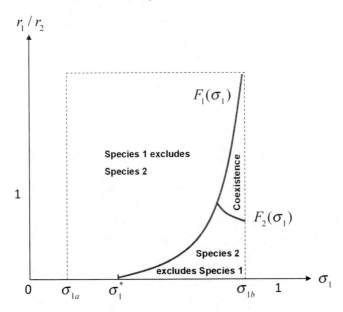

FIGURE 4.13: Diagram for the possibility of invasion or coexistence of plant species determined by plant growth (r_i) and herbivore browsing preference (σ_i).

subscript T is for toxin), where

$$P_{1T}^* = \frac{2\left(G_1 - \sqrt{G_1(G_1 - d/B_1)}\right)}{e_1\sigma_1\left[1 - 2h_1\left(G_1 - \sqrt{G_1(G_1 - d/B_1)}\right)\right]}$$ (4.32)

$$H_T^* = \frac{r_1(1 - P_{1T}^*/K_1)}{e_1\sigma_1/(1 + h_1e_1\sigma_1P_{1T}^*)\left[1 - f_1(P_{1T}^*, 0)/(4G_1)\right]}.$$

It can be verified that $P_{1T}^* > 0$ if $d < B_1C_1(K_1, 0)$, and that E_{1T}^* is biologically feasible if $P_{1T}^* < K_1$.

The Jacobian matrix at E_{1T}^* has two eigenvalues (either real or complex) with a negative real part, and the third eigenvalue is negative if and only if

$$\frac{r_1}{r_2} > \frac{e_1\sigma_1(1 - c_{21}P_{1T}^*/K_2)}{e_2\sigma_2(1 - P_{1T}^*/K_1)}\left(1 - \frac{e_1\sigma_1P_{1T}^*}{4G_1(1 + h_1e_1\sigma_1P_{1T}^*)}\right).$$ (4.33)

It follows that the equilibrium E_T^* is locally asymptotically stable if the condition (4.33) holds. If the inequality (4.33) is reversed then E_T^* becomes unstable, in which case the invasion of species 2 is expected.

Notice that $P_{1T}^* = P_{1T}^*(G_1)$ is a function of G_1 (and that it does not depend on G_2). Thus, the right-hand side of (4.33) defines a function of G_1, which is denoted by $L(G_1)$. It can be shown that $L(G_1)$ is an increasing function of G_1. In the region where r_1/r_2 is above the curve $L(G_1)$, species 2 will be excluded if it starts at a low density, whereas in the region below the curve it is possible for species 2 to invade.

4.2.4   Adaptive foraging by adjusting feeding effort $\sigma_i(t)$

When herbivores forage adaptively by adjusting their feeding effort, $\sigma_1(t)$ and $\sigma_2(t)$, the plant-herbivore dynamics can be very different. Assume that the herbivore feeds optimally at each instant (obtaining the maximum possible energy intake rate) by varying $\sigma_1(t)$ and $\sigma_2(t)$ through time, depending on which selection optimizes the herbivore's immediate growth rate, as long as $\sigma_1(t) + \sigma_2(t) \leq 1$. Therefore, the herbivore can change its feeding strategy rapidly by adjusting its foraging to maximize its overall fitness.

We can compare the outcomes of the models with Holling type 2 functional response and with toxin-determined functional response. When the effect of toxins is removed in (4.28), the TDFRM reduces to a well-known system with a HT2FR. In this case, the outcome under adaptive feeding is demonstrated in Figure 4.14. Although the dynamic behavior shown in Figure 4.14 is in agreement with earlier modeling studies (e.g., [269]) that adaptive feeding can promote plant species coexistence, we observe from Figure 4.14b the complete switching of feeding effort in response to changes in relative abundance of plants. Such behavior would be impossible to achieve in the real world. However, even a small tendency toward frequency-dependent consumption will tend to protect the rarer species. This is very different from the outcomes of the TDFRM, as illustrated next.

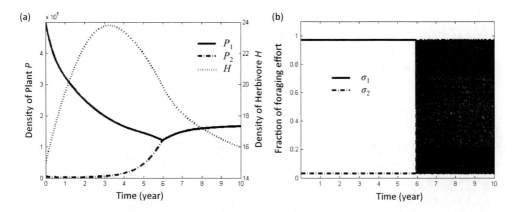

FIGURE 4.14: Simulations of the 3-D model with Holling type 2 functional response for the case of optimal feeding with $\sigma_1 + \sigma_2 \leq 1$. It shows that the two species coexist. It can be observed that in the beginning $\sigma_1 = 1$ and $\sigma_2 = 0$, because the initial density of species 1 is higher and species 1 is decreasing. Once the density of species 2 exceeds that of species 1, the consumption constants switch to $\sigma_1 = 0$ and $\sigma_2 = 1$, and the switches continue to occur (the switches are so frequent that it appears as a black area in the plot σ versus t).

Figure 4.15 illustrates the outcomes of the 3-dimensional TDFRM (4.28) when toxins are present. It shows that, when the herbivore is able to vary its foraging effort through time to maximize fitness, species 2, starting from a low density, cannot invade if G_1 is sufficiently small (i.e., species 1 is sufficiently toxic). Here, although the initial ratio of the invader to resident is very small, $P_2/P_1 = 0.01$, the foraging effort remains greater on species 2 than on species 1. The population of species 2 is pushed rapidly toward extinction (Figure 4.15b). Notice that the relative feeding effort on species 2, $\sigma_2(t)$, remains high even as the population P_2 rapidly declines toward zero (Figure 4.15c). This is in contrast to the model of adaptive herbivory with no plant toxin effects, in which the herbivore switches back and forth to maintain its effort on the more common plant species (Figure 4.14b).

The simulation result illustrated in Figure 4.15 supports the Hypothesis 1 that when plant toxicity is considered, under some conditions adaptive herbivory can result in depensatory effects, favoring exclusion of the less abundant species.

The results shown in Figure 4.15 can change substantially, however, if species 1 is not too toxic (G_1 is not too small). In such cases, avoidance of toxin in species 1 is not so critically

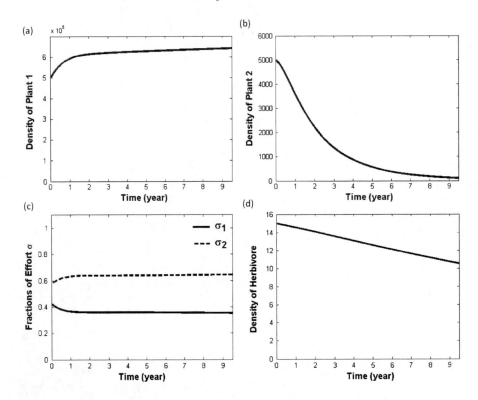

FIGURE 4.15: Simulation results of the TDFRM (4.28) when $G_1 = 35$ and $G_2 = 60$ (species 1 is more toxic). Other parameter values are: $c_{12} = c_{21} = 0.9, r_1 = r_2 = 0.007, K_1 = K_2 = 7 \times 10^5, B_1 = B_2 = 3.4 \times 10^{-5}, e_1 = e_2 = 0.0007, h_1 = h_2 = 0.008, d = 0.0013$. Initial plant densities are $P_{1,0} = 10^5$ and $P_{2,0} = 10^3$.

important for the herbivore, and its efforts can be more strongly determined by the relative population abundances of the two plant species. Some specific scenarios are considered. In one scenario, $G_1 = 50$ and $G_2 = 35$ (more toxic, Figure 4.16a) and it is assumed that $r_1 > r_2$ so that species 1 grows faster than species 2. Other parameter values are the same as in Figure 4.15. In this case, species 2 is able to both invade and achieve higher densities than species 1, by virtue of its higher toxicity. In fact, species 1 goes to extinction. In another scenario, $G_1 = 110$ and $G_2 = 40$ (Figure 4.16b), and two species were allowed to start growing from very small values. The simulations show that species 1 initially increased to a high value instantaneously on the scale of the figure, but was eventually pushed to lower values by the slow growing, heavily defended species 2. Although in this case both plant species persisted, it can be shown that in other cases species 1 will go to extinction.

This result corroborates the Hypothesis 2 that the less defended fast-growing species can be replaced over time by more defended slower-growing species. Natural history observations and experiments on winter foraging of snowshoe hare and moose tend to confirm these results. The per capita daily intake rates of browse (woody twigs, bark) of both species are largely determined by browse toxins (reviewed by Bryant and others [42, 43]). In general, these herbivores feed selectively on rapidly growing early successional species (usually deciduous), show moderate avoidance of mid-successional species, and show strongest avoidance of late successional evergreens such as spruce (*Picea spp.*) and Labrador tea (*Ledum spp.*). Accordingly, consumption of the mid-successional species such as the alders (*Alnus spp.*) is much lower than the intake of early successional species, for example, willow (*Salix spp.*),

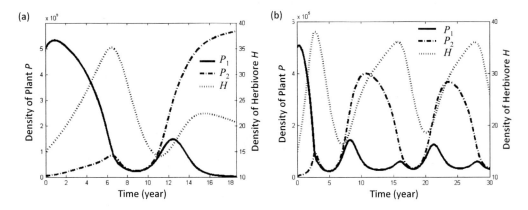

FIGURE 4.16: Simulation results for the 3-D TDFRM for an adaptively foraging herbivore when species 1 is assumed to have a higher growth rate; $r_1 = 0.009 > r_2 = 0.007$, but species 2 is more toxic; $G_1 = 50 > G_2 = 35$ (in (a)) and $G_1 = 110 > G_2 = 40$ (in (b)).

and slightly higher than the intake of late successional evergreens [38, 46, 51, 190, 192]. These observations led to the resource availability hypothesis of plant defense-slow-growing species are more defended than fast-growing species [38, 70]. Kielland and others [192] summarize the browsing research, which focused on the willow-alder transition occurring early in primary succession on the Tanana River Floodplain. On the floodplain, recently deposited sediment is initially colonized by willow and balsam poplar with willow being most abundant [62]. Alder (*A. tenuifolia*) subsequently invades the willow thickets, leading to a willow-alder transition. Increasing dominance of alder also initiates a collapse in wildlife's food supply because alder and the later successional evergreens that replace willow are so toxic that they are hardly eaten [38, 42, 46].

Further model studies on species invasion reveal additional aspects of toxin-influenced dynamics. In the next scenario, $G_1 = 50$, $G_2 = 60$ and $c_{12} = c_{21} = 0.9$, the slightly higher toxicity of species 1 keeps it superior in biomass to species 2 (Figures 4.17a and b). In the case shown in Figure 4.17, the herbivore feeding and the relative abundance of the plant species interact in a way that produces oscillations of all species on a time scale of several years. (The biomass values in these plots refer to edible twigs only, not whole plants.) Note that the changes in $\sigma_1(t)$ vs. $\sigma_2(t)$ through time are not large. Recall in the case of herbivory without toxins, the herbivore shifts completely to expending foraging effort on only one species or the other, whereas here there is only a modest change in foraging effort between the plant species. When the competition coefficients are reduced, $c_{12} = c_{21} = 0.5$, and the system stabilizes with both plant species persisting (Figures 4.17c and d).

The effects of relative toxicity levels of the two species on invasion can be summarized. Invasion is more likely when G_1 increases (resident species becomes less toxic) (Figure 4.18a). For a given set of parameter values, there is a threshold toxin level G_1^* such that invasion is possible only for $G_1 > G_1^*$. Simulations (not shown here) also suggest that the threshold value of G_1^* decreases with increasing initial density of species 2 ($P_{2,0}$). That is, the greater the initial density of the invader, the more toxic must be species 1 if invasion is not to occur (Figure 4.18b), as larger numbers reduce the probability of depensatory herbivory on the invader. Moreover, for each given $G_1 > G_1^*$, there is a range for the ratio G_2/G_1 in which invasion is possible (the upper and lower bounds in the specified region in Figure 4.18a). The results suggest that the ratio G_2/G_1 (relative toxicity) is an important determinant of invasion success for an inferior competitor, and the probability of successful

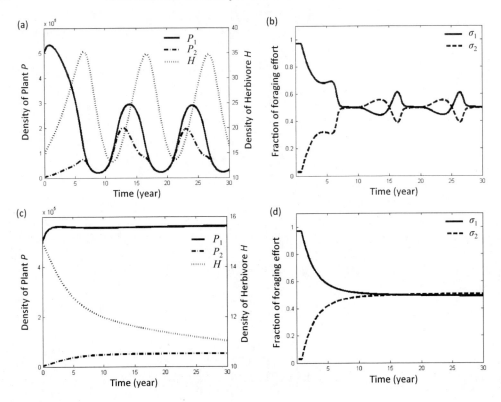

FIGURE 4.17: Similar to Figure 4.15 but toxicity of the resident species 1 is reduced (G_1 changes from 30 to 50). The competition coefficients in (c) and (d) ($c_{12} = c_{21} = 0.5$) are reduced from those in (a) and (b) ($c_{12} = c_{21} = 0.9$), which seems to stabilize the system.

invasion increases with the initial population size of the introduced species. These model results support the Hypothesis 3 that the occurrence of toxin in the resident species and the relative amounts of toxins in the resident and invader can determine the ability of an introduced species to successfully invade in a system with herbivory.

The rare plant effect

As in the case where no plant toxins are involved, it is unlikely that herbivores in the real world would be able to adjust their feeding perfectly in response to toxins to maximize fitness, but some tendencies in this direction are revealed by empirical research. The experimental study by Bergvall et al. [27] with fallow deer (*Dama dama*) supports this model prediction. The authors demonstrated that when the less defended plant (represented by low tannin pellets) was associated with more defended plants (represented by high tannin pellets), the less defended plant was eaten more than if it was growing among its own kind. Even if a less abundant invader is relatively toxic, switching from a highly toxic resident to the invader is still predicted to occur if the invader's toxins and those of the more abundant resident are non-additive. The result is also consistent with Heady's [143] conclusion: "Observations indicate that some species (e.g., the toxic monoterpene containing big sage *Artemesia tridentata*; [346]) are grazed heavily (by livestock) when they occur in small quantities throughout a "better" forage, whereas in dense stands the use is light." This toxin-mediated depensatory effect imposes relatively stronger herbivory on the less abundant invader, thereby reducing its survival, and in the process, reducing plant species

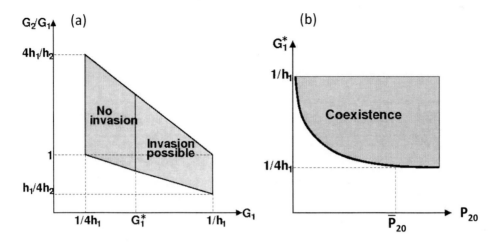

FIGURE 4.18: (a) Range of toxicity values over which an invader (P_2) can succeed when the resident species (P_1) is at equilibrium with the herbivore. It illustrates that the toxicity level of the resident species G_1 can affect the invasion. The shaded regions denote the biologically feasible ranges of G_1 and G_1/G_2. (b) Dependence of coexistence of two species on the population size of invading species and the level of toxicity of the resident species. The threshold value G_1^* decreases with the initial density of species 2 ($P_{2,0}$).

diversity. This depensatory process has be termed the "rare plant effect." Subsequent studies on white-tailed deer (*Odocoileus virginianus*; [17]) and moose (*Alces alces*; [17]) support the prediction.

General predictions from the 3-D TDFRM (4.28)

The TDFRM makes three general predictions. The first is that under conditions of substantial adaptive herbivory, high toxin levels in a plant strongly favor its dominance, in terms of higher biomass levels in the community, over species with lower levels of toxicity. This can be true even though herbivory is often considered to occur in a positive frequency-dependent way; that is, focused on the species that is more abundant to a greater extent than would be expected from random encounters. Instead, when an abundant species is highly toxic, the fraction of effort that is directed at a less abundant or rare species, $P_2(t)$, may remain disproportionately high even though the population biomass of that species is declining rapidly toward zero. The second general prediction is that, in a community of fast-growing species with low toxicity and slow-growing species with high toxicity, and in which there is also significant herbivory, the fast-growing species will dominate in biomass initially, but will be outcompeted by the high-toxicity species in the long run. The third general prediction is that, when the resident species of a community are highly toxic to herbivores, invasion by new species with low toxicity or nonadditive toxins will be difficult, but highly toxic species may be able to invade. All of these predictions appear to be confirmed by empirical data.

The effects of additive plant toxins

In the extreme alternative to non-interactive toxins, that is, the case in which the toxins are the same, or are additive, a different toxin-dependent functional response may be used, denoted by $\tilde{C}(P_1, P_2)$, where

$$\tilde{C}_i(P_1, P_2) = f_i(P_1, P_2)\left(1 - \sum_{i=1}^{2} \frac{f_i(P_1, P_2)}{4G_i}\right), \quad i = 1, 2, \tag{4.34}$$

and f_1 and f_2 are the functions given in(4.30).

To examine this alternative limiting case, the functional response given by equation (4.34) is substituted into equations in (4.28) for $C_i(P_1, P_2)$. In this case, the ingestion of a toxin from one plant species negatively affects the feeding rate on both species. Only a few results are presented here to indicate the general trend that inclusion of an interactive effect of the toxins has on the dynamics. For example, Figure 4.19 shows the dynamics in the case when the functions $\tilde{C}_i(P_1, P_2)$ in (4.34) are used in the model (4.28).

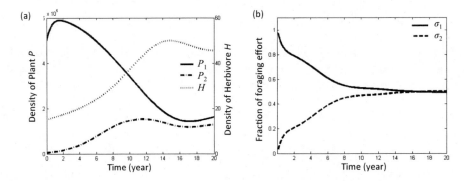

FIGURE 4.19: Similar to Figure 4.15 but it is for the case when the non-interactive functional response $C_i(P_1, P_2)$ is replaced by $\tilde{C}_i(P_1, P_2)$ in the model (4.28).

In Figure 4.15, the resident species excluded the rare, less toxic invading species. When the functional response $\tilde{C}(P_1, P_2)$ in equation (4.34) is substituted in the model (4.28), however, the two species are able to coexist, as illustrated in Figure 4.19. Thus, the additive toxins do not show the same effect of non-interacting toxins in reversing the effect of herbivore compensation on the competing plants. Instead, the situation is tilted toward herbivore facilitated coexistence.

4.3 Plant-herbivore interactions with age-dependent toxicity

In this section, a model with age-dependent toxicity is considered. The results are presented in [227]. An added complication in taking into account PSMs in plant biomass is that the concentrations can vary according to the specific biomass component of the plant, and even within biomass components of the same component type, but differing in age. In this section the plant tissues of interest are the segments of twigs of deciduous woody plants ranging in age from younger to older. The younger segments of twigs (YST) are the tips of twigs and the older segments of twigs (OST) are the segments supporting the twig tips.

In all winters, except the winters during the hare crash and the early low phase of the hare cycle, twigs make up more than 99% of a snowshoe hare's diet. There is no previous study that has even tried to estimate the amount of bark that snowshoe hares eat, because most people do not consider bark to be an important source of food. Good examples of the importance placed on twigs are the chapter on shrubs in the report on the Kluane Project at Kluane National Park, Yukon Territory, Canada, edited by Krebs et al. [200] and Pease et al.'s [290] excellent study done in northern Alberta, Canada.

The diet of the snowshoe hares is in fact even narrower, as the diet is in many locations largely restricted to twigs of only a few woody plant species, such as *Salix glauca* and *Betula glandulosa* at Kluane National Park. The tight coupling of the hare diet to only a few species is especially important in view of possible further limitation due to toxins in twig biomass, as it implies a vulnerability to food limitation at times. Even when there is an abundance of twigs of a preferred species (e.g., grayleaf willow *S. glauca* at Kluane), twig biomass can be insufficient to support hares. This is because toxic secondary metabolites limit the intake of any one species to below the per capita daily browse requirement of a hare. A good example of this multi-species diet requirement is in the paper by Rodgers and Sinclair [314]. They found (their Table 1) that the intake of the most preferred of the preferred browses (*S. glauca* and *B. glandulosa*) was less than 50 g dry mass/kg hare/day. Even in warm winters the per capita daily intake of high-quality browse (low-toxicity twigs less than 3–4 mm diameter) required by a snowshoe hare is about 100 g dry mass/kg hare/day [30], and in colder periods the required intake increases to about 120 g dry mass/kg hare/day (Reichardt et al. [307]). Thus, Kluane hares need to feed on the twigs of more than one preferred species to survive, and these twigs are predominantly the twigs of shrubs. In Krebs et al. [200] book it is stated that about 98% of the available shrub twig biomass at Kluane is gray willow (*S. glauca*). In the winter of a hare crash at Kluane hares ate almost all of the available twigs of *B. glandulosa* (Turkington et al. [362]), which is the next most abundant (Krebs et al. [200]) preferred browse species at Kluane [314]. Also, Melnychuk and Krebs ([255], p. 403) state: "Birch shrubs are the preferred winter food for hares and are heavily browsed during the hare peak, with hares typically eating 90% of all birch branches at the peak in contrast to only 20–30% of willow branches." Thus, in the years of the hare crash and the early low phase of the hare cycle at Kluane, hares would very likely have been food limited.

4.3.1 The model with age-dependent toxicity

Assume that the twigs of the plant are split into two age stages, YST and OST. YST consists of all the twig biomass extending from newly formed biomass up to biomass of a specified older age τ_1, usually at least a year, while OST corresponds to all older twig biomass that is still small enough in diameter to be consumed by the herbivore. Let $T_1(t)$ denote the total twig biomass of YST (twig ages between $(0, \tau_1)$), and let $T_2(t)$ denote the total twig biomass of OST, $T_2(t)$ (twig ages between (τ_1, τ_2)). The biomass of herbivore is denoted by $P(t)$. The following model (acronym TSM, which stands for twig segment model) is developed and analyzed in [227]:

$$T_1(t) = \int_0^{\tau_1} b(T_1(t-a)) e^{-\lambda_1 a} \exp\left(-\int_{t-a}^t \left(\rho_1(s) + \rho_2(s)\right) P(s) ds\right) da,$$

$$T_2(t) = \int_{\tau_1}^{\tau_2} b(T_1(t-a)) \exp\left(-\int_{t-a}^{\tau_1+t-a} \left[\lambda_1 + \left(\rho_1(s) + \rho_2(s)\right) P(s)\right] ds \right. \tag{4.35}$$
$$\left. -\int_{\tau_1+t-a}^t \left[\lambda_2 + \rho_2(s) P(s)\right] ds\right) da,$$

$$p'(t) = B_2 \rho_2(t) T_2(t) P(t) - dP(t),$$

where $b(\cdot)$ is the birth rate of new twigs, which is a decreasing function of its argument; λ_i $(i = 1, 2)$ denote the natural per-capita twig loss for twigs in age stage i; B_2 is the conversion factor between plant biomass and the number of herbivores, and

$$
\rho_1(t) = \frac{e_1\sigma_1}{1 + e_1\sigma_1\left(h_1 + \frac{1}{4G_1}\right)T_1(t) + e_2\sigma_2\left(h_2 + \frac{1}{4G_2}\right)T_2(t)},
$$

$$
\rho_2(t) = \frac{e_2\sigma_2}{1 + e_1\sigma_1\left(h_1 + \frac{1}{4G_1}\right)T_1(t) + e_2\sigma_2\left(h_2 + \frac{1}{4G_2}\right)T_2(t)}.
$$

(4.36)

Other parameters are defined in Table 4.3.

Note that the form for toxin-dependent function response ρ_i given in (4.36) is an approximation of the function C_i described in (4.29), as analogous to the 2-D model discussed in Section 4.1.1, i.e., the alternative function response (4.4) is an approximation of the functional form given in (4.2) and (4.3).

In the T_1 equation of the system (4.35), the first term on the right-hand side represents twig loss, either natural or because of herbivory. Herbivory of a twig of age in $(0, \tau_1)$ could be direct (i.e., an herbivore was explicitly looking for a twig of this age stage) or indirect (i.e., the unfortunate young twig happened to be a segment of an older twig of age in (τ_1, τ_2), and the herbivore wanted the latter). The last term represents the rate at which twigs leave the T_1 and enter the T_2 stage, having survived all possible forms of death. This term is the birth rate at the earlier time $t - \tau_1$ multiplied by the probability of survival to age τ_1.

In the T_2 equation of (4.35), the last term on the right-hand side is the rate at which twigs exit from the T_2 stage (they remain on the plant but are now major branches too large to eat). This happens τ_2 time units after birth, hence the presence of the factor $b(T_1(t - \tau_2))$, and each exponential factor represents a survival probability. Those involving λ_1 and λ_2 represent the probabilities of twig segments surviving natural mortality during the first and second stages of life, respectively, for which the terms YST and OST are introduced, respectively, above. The exponential factor involving $\rho_1(s) + \rho_2(s)$ in the integrand is the probability of surviving herbivory during the first stage, which could be direct or indirect. A segment exiting the T_2 stage at time t was in the first stage of its life between times $t - \tau_2$ and $t - \tau_2 + \tau_1$ and during this time it had to survive both direct herbivory by herbivores who were searching for segments of this age stage, and also indirect herbivory by herbivores who were really searching for older segments of age in (τ_1, τ_2) but necessarily removed the younger segments of such twigs as well. The final exponential factor is the probability that the segment survived its second stage (from time $t - \tau_2 + \tau_1$ to time t) during which stage it again became vulnerable to herbivores searching for twigs in the (τ_1, τ_2) age stage.

The initial conditions have the following form:

$$
T_1(s) = T_1^0(s) \geq 0, \quad T_2(s) = T_2^0(s) \geq 0, \quad P(s) = P^0(s) \geq 0, \quad s \in [-\tau_2, 0];
$$

$$
T_1^0(t) = \int_0^{\tau_1} b(T_1^0(-a))e^{-\lambda_1 a} \exp\left(-\int_{-a}^0 (\rho_1^0(s) + \rho_2^0(s))P^0(s)ds\right)da,
$$

$$
T_2^0(t) = \int_{\tau_1}^{\tau_2} b(T_1(-a)) \exp\left(-\int_{-a}^{\tau_1-a} \left[\lambda_1 + (\rho_1^0(s) + \rho_2^0(s))P^0(s)\right]ds\right.
$$

(4.37)

$$
\left. -\int_{\tau_1-a}^0 \left[\lambda_2 + \rho_2^0(s)P^0(s)\right]ds\right)da,
$$

where $T_1^0(s)$, $T_2^0(s)$, and $P^0(s)$ are prescribed functions, and ρ_1^0 and ρ_2^0 are given by expressions of the same form as (4.36) but with zero superscripts on T_1 and T_2.

TABLE 4.3: Definition of parameters and values used in simulations.

Parameter	Definition	Value
λ_i	Natural per-capita segments loss rate	0
e_i	Rate of encounter per unit plant biomass T_i, $i = 1, 2$	0.0007
σ_1	Selection coefficient of T_1, $0 \leq \sigma_1 \leq 1$	0.1
σ_2	Selection coefficient of T_2, $0 \leq \sigma_2 \leq 1$	0.9
h_i	Time for handling one unit of the plant T_i, $i = 1, 2$	0.008
G_1	Measure of plant toxicity for the YST T_1	1
G_2	Measure of plant toxicity for the OST T_2	100
r	Growth rate of YST	1
K	Carrying capacity of new YST	10000
d	Per-capita death rate of the herbivore unrelated to plant toxicity	1/365
B_2	Conversion constant (herbivore biomass per unit of plant T_2	0.0017
τ_1	Duration of YST (effectively inedible) stage	365×3
τ_2	Age beyond which twigs are too large to eat	365×3

4.3.2 Herbivore extinction and oscillations

Assume that $b(\cdot) > 0$ is a decreasing function. The system (4.35) has a unique hare-extinction equilibrium $E^* = (T_1^*, T_2^*, 0)$, where $T_1^* > 0$ and $T_2^* > 0$ satisfy the following equations

$$\lambda_1 T_1^* = b(T_1^*)(1 - e^{-\lambda_1 \tau_1}),$$

$$\lambda_2 T_2^* = b(T_1^*)e^{-\lambda_1 \tau_1}(1 - e^{-\lambda_2(\tau_2 - \tau_1)}).$$

(4.38)

It is shown in [227] that under a certain condition (particularly when the hare mortality d is sufficiently high), the equilibrium E^* is globally asymptotically stable.

When E^* is unstable, particularly when τ_1 is not too large, other dynamics including persistence of all three variables (T_1, T_2, and P) in the form of limit cycle oscillation are possible. The dependence of the model behavior on various factors such as the delay τ_i and plant toxicity G_i ($i = 1, 2$) can be explored numerically. This is demonstrated in Figure 4.20. For these simulations, the functional form for b is chosen to be $b(T_1) = r\exp(-T_1/K)$, where r is the growth rate of new segments of twigs and K is the carrying capacity of the new segments. Other parameters are listed in Table 4.3.

The first row of Figure 4.20 shows that increasing the value of τ_1, which corresponds to an increasing age at which the twigs switch from toxic YST to lower-toxicity OST, increases the limit cycle period, as well as pushing the system toward the Hopf bifurcation threshold, where the limit cycle oscillations become stabilized. Further increases in τ_1 ultimately lead to extinction of the herbivore (see the first row in Figure 4.20). This is due to the decreasing amount of edible biomass with increasing τ_1. Increases in the plant growth rate, r, from 0.5 to 2 (see the second row in Figure 4.20), cause the interaction to change from stability to large-amplitude limit cycle oscillations. This appears to be similar to the well-known "paradox of enrichment," in which increases in production of the prey lead to instability in a predator-prey interaction. The influence of G_1 (toxicity of the YST of the twig) is shown in the third row of Figure 4.20. The increased damping rate with increasing G_1 is consistent with what one would expect from increasing total handling time of the plant biomass. The term $1/G_1$ acts as a handling time, because the rate at which the herbivore can detoxify toxin sets a limit on intake rate. This is a result of the Michaelis–Menten kinetics of detoxification [78], and thus $1/G_1$ is an analog of the handling time, h, in the Holling type 2 functional response [158, 159]. Note that for high enough toxicity ($G_1 = 0.1$) the herbivore dies out. Changes in G_2 do not have much effect on the dynamics (see the fourth row in Figure 4.20), because the proportion of contribution of G_2 to the handling time is small, for the range of values of G_2 that are considered above. Numerical simulations shown in Figure 4.20 demonstrate the existence of limit cycles over ranges of parameters reasonable for hares browsing on woody vegetation in boreal ecosystems. This showed that age dependence in plant chemical defenses has the capacity to cause hare-plant population cycles.

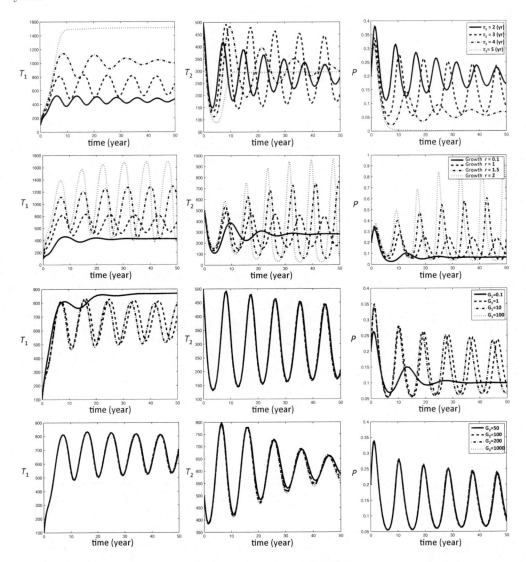

FIGURE 4.20: Numerical simulations of model (4.35) showing the dependence of model behavior on age τ_1 (row 1), growth r (row 2), toxicity levels G_1 (row 3) and G_2 (row 4). Numerical simulations illustrating the existence of limit cycles over ranges of parameters reasonable for hares browsing on woody vegetation in boreal ecosystems.

A basic property that emerged through the variation in twig toxicity with age is a time lag that can generate plant-herbivore limit cycle behavior over certain parameter ranges. This is of interest, because oscillatory dynamics are a common feature of the dynamics of snowshoe hare populations in boreal ecosystems. Although many mechanisms have been offered to explain such cycles [31, 36, 45, 101, 107, 108, 131, 187, 188, 200, 290, 329], the fact that the changes in toxicity with age in the edible parts of plants can easily produce the sort of time lags that can generate cycles is of some interest and should be explored. The numerical simulations in this study show many examples of both stable limit cycles and damped oscillations. Although it is not clear which parameter values may apply to a given situation in nature, it is believed that parameter ranges that produce limit cycles in this model are feasible ones for at least some natural situations.

The dependence of the period of oscillation in the model on τ_1 is of special interest, because time delays are a common cause of oscillations. The rough estimates of τ_1 being about 2 years produces a cycle of about 5 or 6 years. This is slightly shorter than the snowshoe hare cycle, which has a period which ranges from 7 to 17 years [131]. This minor disagreement of the model with empirical estimates of the cycle suggests that plant defense is not the total explanation. One possible argument that plant defense is at least a partial explanation for cycling is that a good case can be made for increasing the value of τ_1 used in the model. The model predicts that if the restoration of less defended OST takes 3–4 years, then hare numbers will be suppressed for a sufficiently long period of time after the hare population crashes to generate a hare cycle period within the range often observed in nature, 7–17 years. This OST retardation of 3–4 years has been measured in feeding experiments with Alaskan snowshoe hares using several woody species that are, before severe browsing, preferred winter foods of Alaskan snowshoe hares [108]. Experiments with Wisconsin quaking aspen also suggest that severe browsing could increase twig toxicity for 2–5 years [86]. Alternatively, other factors may exist, such as predation [188] and stress of the hare caused by fear of predation [31, 329] may serve the same purpose of prolonging the low numbers of hares following a crash.

4.4 Reaction-diffusion modeling approach

Climate change is apparently already causing latitudinal and altitudinal shifts in ecosystems, as a result of changes in temperature and precipitation over the past few decades, as reviewed in [377]. The upward shift in temperatures in some places is creating conditions favorable to the invasion of vegetation that is adapted to warmer temperatures. Ecotones are places where the effects of climate changes are most likely to be evident [265], and one type of ecotone at which changes in vegetation have been observed both latitudinally and altitudinally is that between woody vegetation and tundra or Alpine sedges, grasses and mosses, where woody shrubs have advanced in places; see [263] for documentation in numerous sites. Woody vegetation, either in prostrate or erect form, may have an advantage over graminoids and other non-woody plants as temperatures increase. An important question is whether the woody plant–tundra ecotone will advance at a pace set by climate change, or whether biological factors will also affect this rate. In addition to shifts in ecotones, shifts in community composition may occur within biomes, as species adapted to higher temperatures already present increase in abundance or new ones invade. For example, the boreal forest is being invaded by more temperate species, changing its composition [102], and alpine grassland plant communities may appear to be changing in composition [32]. However, there is evidence that climatic factors alone may not determine the changes that occur. In particular, herbivory by small and large mammals may slow the advance [102, 181, 277, 294, 339]. Post and Pedersen [294] noted that caribou and reindeer can constrain biomass of deciduous shrubs such as dwarf birch and willow, and perhaps slow their advance into tundra. Speed et al. [339] report on the effect of sheep in slowing the advance of alpine grassland communities, or even causing a downslope shift. Modeling has been used to describe possible interaction of temperature change and grazing rate on a tundra plant community is Siberia, showing a shift away from deciduous shrub toward graminoid and moss tundra in some cases, but toward less palatable evergreen species in others [388].

The apparent importance of herbivory in relation to vegetation changes accompanying climate change indicates that it is also important to obtain a better understanding of how the level of plant chemical defense affects herbivory, and thus, indirectly, the relationship between climate and the invasion of new plant species in a community. This spatial modeling work is based on the knowledge from field studies and modeling that plant toxins can have a significant impact on the outcomes of plant-herbivore interactions (see, for example, [62, 97, 98, 99, 190, 192]). These models have been used to test the hypothesis that toxin-dependent selective herbivory by mammals has resulted in invasion of more toxic plants during plant succession in taiga forest [37, 51, 95, 190, 192], the southern boreal forest [287], and temperate grasslands [66, 67].

In all of the TDFRM studied in the previous sections, the spatial component was ignored and the models were systems of ODEs. In this section the spatial extension of the TDFRM model is considered by taking into consideration the herbivore movement [96]. In this model, the plant species with a higher level of toxicity is assumed to be less preferred by the herbivore and to have a relatively lower intrinsic growth rate than the less toxic plant species. Two of the equilibrium points of the system representing significant ecological interests are E_1, in which only the less toxic plant is present, and E_2, in which the more toxic plant and herbivore coexist while the less toxic plant has gone to extinction. Under certain conditions it is shown that, for the spatially homogeneous system all solutions will converge to the equilibrium E_2, whereas for the reaction-diffusion model there exist traveling wave solutions connecting E_1 and E_2. This has important implications for the invasion of toxic species.

4.4.1   A 2-D TDFRM with herbivore movement included

Assume that the movement is random, which can be modeled as a diffusion process. The model considered in this section can be used to identify conditions under which the model has a traveling wave solution, which connects the equilibrium where only plant species 1 is present ($P_1 > 0$) to another equilibrium at which $P_1 = 0$ while the more toxic plant and the herbivore coexist. The reaction-diffusion model in this section is a spatial extension of the 3-D TDFRM (4.28) considered in Section 4.2.1, with a simplification that the toxicity for plant species 1 is not explicitly included in the equations. The model reads (for ease of presentation different notations have been used for some of the parameters):

$$\frac{dP_i}{dt} = \mathcal{A}_i(P_1, P_2) - C_i(P_1, P_2)H, \quad i, j = 1, 2, \ i \neq j,$$

$$\frac{dH}{dt} = \sum_{i=1}^{2} B_i C_i(P_1, P_2)H - dH, \tag{4.39}$$

where

$$\mathcal{A}_i(P_1, P_2) = r_i P_i \left(1 - \frac{P_i + c_{ij} P_j}{K_i} \right), \quad i = 1, 2 \tag{4.40}$$

and

$$C_1(P_1, P_2) = \frac{\sigma_1 P_1}{\mathcal{F}(P_1, P_2)},$$

$$C_2(P_1, P_2) = \frac{\sigma_2 P_2}{\mathcal{F}(P_1, P_2)} \left(1 - \frac{\sigma_2 P_2}{4G\mathcal{F}(P_1, P_2)} \right) \tag{4.41}$$

with

$$\mathcal{F}(P_1, P_2) = 1 + \sigma_1 h_1 P_1 + \sigma_2 h_2 P_2. \tag{4.42}$$

Assume that the movement (or seed dispersal) of the plant can be neglected on the short time scale considered here. Then, the model (4.39) can be extended to the following reaction-diffusion system

$$\frac{\partial P_i}{\partial t} = \mathcal{A}_i(P_1, P_2) - C_i(P_1, P_2)H, \qquad i = 1, 2,$$
$$\frac{\partial H}{\partial t} = \rho \Delta H + \left(B_1 C_1(P_1, P_2) + B_2 C_2(P_1, P_2) - d \right)H, \tag{4.43}$$

where $P_i(t, x)$ and $H(t, x)$ are population densities of plants and herbivore, respectively, at time t and location $x \in \mathbf{R}^2$; ρ denotes the diffusivity of herbivore; $\Delta = \sum\limits_{i=1}^{2} \partial^2/\partial x_i^2$ is the Laplace operator; $\mathcal{A}_i$ and C_i are defined in (4.40) and (4.41).

As the plant species 2 (P_2) is more toxic than plant species 1 (P_1), the following conditions are biologically reasonable and assumed to hold in this section:

(I) $\sigma_1 > \sigma_2$, which represents the fact that the toxic plant species is less preferred by the herbivore;

(II) $r_1 > r_2$, which implies that the toxic plant species has a reduced intrinsic growth rate as a trade-off of cost for its investment in defense against herbivore browsing.

For ease of notation, we normalize K_1 and K_2 in system (4.39) to let

$$K_1 = K_2 = 1, \tag{4.44}$$

and let

$$C(P) = C_2(0, P), \quad g(P) = g_2(0, P).$$

System (4.39) has several possible equilibria. However, in this section only the following two equilibria are considered:

$$E_1 = (1, 0, 0) \quad \text{and} \quad E_2 = (0, P_2^*, H^*),$$

where $0 < P_2^* \leq 1$ and $H^* > 0$. E_1 is the equilibrium at which only plant species 1 is present, whereas at E_2 the plant species 2 and the herbivore coexist with the plant species 1 being absent. The reason that only these two equilibria are considered is because the main biological question to be considered here is how herbivore browsing may influence the invasion of the toxic plant in space, and the approach used here to address this question is by exploring the existence of traveling wave solutions connecting from E_1 to E_2. To this end, additional assumptions are needed. Let

$$g_i(P_1, P_2) = \frac{\mathcal{A}_i(P_1, P_2)/P_i}{C_i(P_1, P_2)/P_i}$$

$$= \begin{cases} \dfrac{r_1 \left[1 - (P_1 + c_{12}P_2)/K_1 \right]}{\sigma_1/\mathcal{F}(P_1, P_2)}, & i = 1, \\[4mm] \dfrac{r_2 \left[1 - (P_2 + c_{21}P_1)/K_2 \right]}{\dfrac{\sigma_2}{\mathcal{F}(P_1, P_2)} \left[1 - \dfrac{\sigma_2 P_2}{4G\mathcal{F}(P_1, P_2)} \right]}, & i = 2, \end{cases} \tag{4.45}$$

where $\mathcal{A}_i$, C_i, $i = 1, 2$, and F are given in (4.40)–(4.42). Note that the numerator $\mathcal{A}_i(P_1, P_2)/P_i$ is the per-capita growth rate of plant species i, and the denominator $C_i(P_1, P_2)/P_i$ is the per-capita consumption rate of plant i per herbivore. Thus, g_1 and g_2 represent the ratios of the growth to loss of plant species 1 and 2, respectively, which can be used as a measure for the overall competitive ability (or fitness) of the plant.

For the purpose of studying the invasion of a plant species (e.g., species 2) via the existence of traveling wave solutions in the spatial extension of the model (4.39), only parameter regions in which specific properties of the system hold will be considered. This requires the following assumptions:

A1. $g_1(P_1, P_2) > g_2(P_1, P_2)$.

A2.   There is a unique $P_2^* \in (0, 1)$ such that $B_2 C(P_2^*) - d = 0$. This excludes the parameter region in which multiple interior equilibria of the form E_2 exist.

A3. $g'(P_2^*) < 0$. This excludes the parameter region in which a limit cycle exists.

The following assumption on g is related to A3:

A3′. $g(U) > g(P_2^*)$ for $0 \le U < P_2^*$ and $g(U) < g(P_2^*)$ for $P_2^* < U \le 1$.

We now consider the possibility of invasion of plant 2 (P_2) into an environment in which plant 1 (P_1) has established in the absence of plant 2 and herbivore (H). A successful invasion by plant 2 in the presence of the herbivore and extinction of plant 1 can be represented by a transition from the equilibrium $E_1 = (1, 0, 0)$ to $E_2 = (0, P_2^*, H^*)$ in space. The existence and uniqueness of E_2 are guaranteed under Assumption A2. This can be investigated by examining whether there will be a zone of transition from the equilibrium E_1 to the equilibrium E_2. Mathematically, this transition can be described by a particular type of solutions, i.e., the traveling wave solutions of the form

$$P_i(t, x) = U_i(k \cdot x + ct), \quad i = 1, 2,$$
$$H(t, x) = V(k \cdot x + ct) \tag{4.46}$$

under the boundary condition

$$(U_1(-\infty), U_2(-\infty), V(-\infty)) = E_1 = (1, 0, 0),$$
$$(U_1(\infty), U_2(\infty), V(\infty)) = E_2 = (0, P_2^*, H^*). \tag{4.47}$$

The constant c is the wave speed and $k \in \mathbf{R}^n$ is a unit vector denoting the direction of wave propagation. Equation (4.43) has a traveling wave solution of form (4.46) connecting E_1 and E_2 if and only if the functions $U_1(\xi)$, $U_2(\xi)$, and $V(\xi)$, with $\xi = x \cdot k + ct$, form a solution of the system

$$\dot{U}_i = \mathcal{A}_i(U_1, U_2) - C_i(U_1, U_2)V, \quad i = 1, 2,$$
$$c\dot{V} = \rho \ddot{V} + \Big(B_1 C_1(U_1, U_2) + B_2 C_2(U_1, U_2) - d \Big) V. \tag{4.48}$$

Introduce the variable

$$W = cV - \rho \dot{V}.$$

Then the system (4.48) and boundary condition (4.47) are transformed to the equivalent system

$$\dot{U}_i = \mathcal{A}_i(U_1, U_2) - C_i(U_1, U_2)V, \qquad i = 1, 2,$$

$$\dot{V} = \frac{1}{\rho}(cV - W), \tag{4.49}$$

$$\dot{W} = \Big(B_1 C_1(U_1, U_2) + B_2 C_2(U_1, U_2) - d\Big)V$$

with the boundary condition

$$(U_1(-\infty), U_2(-\infty), V(-\infty), W(-\infty)) = E_1^* = (1, 0, 0, 0),$$

$$(U_1(\infty), U_2(\infty), V(\infty), W(\infty)) = E_2^* = (0, P_2^*, H^*, cH^*). \tag{4.50}$$

4.4.2 Existence of traveling wave solutions

A technique developed in [162] can be used to show the existence of solutions to (4.49) satisfying the boundary condition (4.50). Let

$$M = \max\Big\{B_1 C_1(P_1, P_2) + B_2 C_2(P_1, P_2) : 0 \le P_i \le 1, \, i = 1, 2\Big\}. \tag{4.51}$$

For $c > 2\sqrt{\rho(M - d)}$, construct a set $\Omega \subset \mathbf{R}^4$ as follows. Let

$$\Omega = \left\{ \begin{array}{c} (U_1, U_2, V, W): \quad 0 \le U_i \le 1, \, i = 1, 2, \, 0 \le \gamma_1 V \le W \le \gamma_2 V, \\[2mm] B_1 U_1 + B_2 U_2 + W \le \mathcal{H} \end{array} \right\} \tag{4.52}$$

where γ_1, γ_2, and H are constants with

$$\gamma_1 = \frac{c}{2}, \qquad \gamma_2 = \frac{c + \sqrt{c^2 + 4\rho d}}{2}, \tag{4.53}$$

$$\mathcal{H} > \frac{\gamma_2 M}{d} + B_1 + B_2.$$

The boundary of Ω consists of the faces $S_1 - S_8$ with

$$S_1 = \{W = \gamma_1 V > 0, \, 0 < U_i < 1, \, i = 1, 2\},$$

$$S_2 = \{W = \gamma_2 V > 0, \, 0 < U_i < 1, \, i = 1, 2\},$$

$$S_3 = \{B_1 U_1 + B_2 U_2 + W = \mathcal{H}, \, \gamma_1 V \le W \le \gamma_2 V, \, 0 < U_i \le 1, \, i = 1, 2\},$$

$$S_4 = \{U_1 = 0, \, \gamma_1 V \le W \le \gamma_2 V, \, 0 < U_2 < 1, \, i = 1, 2\},$$

$$S_5 = \{U_2 = 0, \, \gamma_1 V \le W \le \gamma_2 V, \, 0 < U_1 < 1, \, i = 1, 2\}, \tag{4.54}$$

$$S_6 = \{U_1 = 1, \, \gamma_1 V \le W \le \gamma_2 V, \, 0 < U_2 < 1, \, i = 1, 2\},$$

$$S_7 = \{U_2 = 1, \, \gamma_1 V \le W \le \gamma_2 V, \, 0 < U_1 < 1, \, i = 1, 2\},$$

$$S_8 = \{V = W = 0, \, 0 \le U_i \le 1.\}.$$

Lemma 4.1. *A solution of (4.49) through a point in Ω can only exit from a point in $S_1 \cup S_2$ (see Figure 4.21).*

Proof: First, a solution cannot exit Ω from a point in $S_4 \cup S_5 \cup S_8$, which is invariant. Also a solution cannot exit Ω from a point in S_6 or S_7 because $\dot{U}_1 < 0$ at S_6 and $\dot{U}_2 < 0$ at S_7. Let us consider the direction of the vector field of (4.49) on S_3. For a point $(U_1, U_2, V, W) \in S_3$,

$$V \geq \frac{W}{\gamma_2},$$

$$W = \mathcal{H} - B_1 U_1 - B_2 U_2 \geq \mathcal{H} - B_1 - B_2 > \frac{M\gamma_2}{d}, \tag{4.55}$$

$$M \geq B_1 \mathcal{A}_1(U_1, U_2) + B_2 \mathcal{A}_2(U_1, U_2).$$

Notice that the out normal vector of the face S_3 is $(B_1, B_2, 0, 1)^T$. Let $Z = (\dot{U}_1, \dot{U}_2, \dot{V}, \dot{W})^T$ be the vector filed of (4.49) at the point $(U_1, U_2, V, W) \in S_3$. Then, the use of (4.49) and (4.55) yields that

$$\begin{aligned} Z \cdot (B_1, B_2, 0, 1)^T &= B_1 \mathcal{A}_1(U_1, U_2) + B_2 \mathcal{A}_2(U_1, U_2) - dV \\ &\leq M - \frac{d}{\gamma_2} W \\ &\leq M - \frac{d}{\gamma_2}(\mathcal{H} - B_1 - B_2) < 0, \end{aligned} \tag{4.56}$$

which implies that the vector field at a point in S_3 points toward the inside of the region Ω.

Next, consider the vector field on the face S_1. The out normal vector of the face S_1 is $(0, 0, \gamma_1, -1)^T$. Then, by (4.49), the definitions of γ_1 and the number M, and the equality

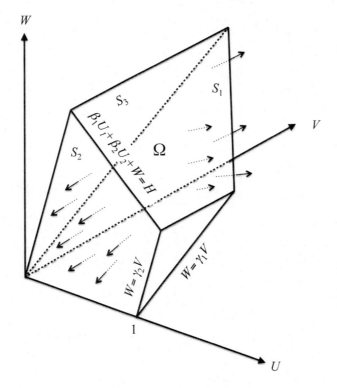

FIGURE 4.21: The region Ω defined in (4.52). In the figure the axis **U** stands for U_1-U_2 plane. A solution can only exit in the region Ω from either a point in the face S_1 or a point in the face S_2.

$W = \gamma_1 V$ for points in S_1, it can be obtained that

$$
\begin{aligned}
Z \cdot (0, 0, \gamma_1, -1)^T &= \frac{\gamma_1}{\rho} \big[cV - W \big] - \big[B_1 C_1(U_1, U_2) + B_2 C_2(U_1, U_2) - d \big] V \\
&\geq \left(\frac{\gamma_1}{\rho} [c - \gamma_1] - [M - d] \right) V \\
&= \frac{1}{4\rho} \big[c^2 - 4\rho(M - d) \big] V \\
&> 0.
\end{aligned}
\tag{4.57}
$$

The inequality (4.57) implies that the vector field at a point in S_1 points toward the outside of the region Ω. Similarly, it can be verified that the vector field at a point in S_2 points toward the outside of the region Ω. This completes the proof.

Lemma 4.2. *Suppose the conditions A1 and A2 are satisfied. Let*

$$
\Psi(t) = (U_1(t), U_2(t), V(t), W(t))
$$

be a solution of (4.49). If

$$
\Psi(t) \in \text{Int}\,\Omega, \quad t \geq 0,
$$

where Int Ω denotes the interior of Ω, then

(a) *$U_1(t) \to 0$ as $t \to \infty$.*

(b) *As $t \to \infty$, $\Psi(t)$ does not converge to the origin $E_0 = (0, 0, 0, 0)$ or the equilibrium point $E_0^* = (0, 1, 0, 0)$ of (4.49).*

Proof: First, $\Psi(t) \in \text{Int}\,\Omega$ implies that $V(t) > 0$ and $0 < U_i(t) < 1$ for $i = 1, 2$. It can be shown that $U_1(t) \to 0$ as $t \to \infty$ (see [96] for a more detailed derivation). Next, it can also be shown that, as $t \to \infty$, $\Psi(t)$ does not converge to the origin as $t \to \infty$. From the definitions of $\mathcal{A}_2(U_1, U_2)$ and $C_2(U_1, U_2)$ it is easy to see that there are positive constants ϵ and δ such that

$$
\frac{1}{U_2} \big[\mathcal{A}_2(U_1, U_2) - C_2(U_1, U_2) V \big] \geq \delta
\tag{4.58}
$$

for $0 \leq V \leq \epsilon$, $0 < U_i \leq \epsilon$, $i = 1, 2$. Suppose $\Psi(t) \to E_0$ as $t \to \infty$. Then there is a t_0 such that $0 \leq V(t) \leq \epsilon$, $0 < U_i(t) \leq \epsilon$, $i = 1, 2$, for all $t \geq t_0$. On the other hand, (4.58) and the second equation of (4.49) imply that for $t \geq t_0$,

$$
\dot{U}_2(t) \geq U_2(t)\delta, \quad t \geq t_0.
$$

The last inequality implies that there must be a time $t_1 > t_0$ such that $U_2(t_1) > \epsilon$, which leads to a contradiction. Finally, A2 implies that $B_1 C_1(0, 1) + B_2 C_2(0, 1) - d > 0$. Hence, there is a small neighborhood $\mathcal{C}$ of $(0, 1) \in \mathbf{R}^2$ such that $B_1 C_1(U_1, U_2) + B_2 C_2(U_1, U_2) - d > 0$ for $(U_1, U_2) \in \mathcal{C}$. This inequality and the equation for W in (4.49) immediately yield that $\Psi(t)$ cannot remain in a sufficiently small neighborhood of E_0^*. The proof is completed.

Lemma 4.2 implies that for a solution $\Psi(t)$ of (4.49), if it stays inside of Ω for $t \geq 0$, then its ω-limit set exists and it is a bounded, connected invariant set of the reduced system of (4.49) for which $U_1 \equiv 0$:

$$\dot{U} = C(U)\left[g(U) - V\right], \qquad i = 1, 2,$$

$$\dot{V} = \frac{1}{\rho}[cV - W], \qquad\qquad\qquad (4.59)$$

$$\dot{W} = [B_2 C(U) - d]\, V,$$

where

$$C(U) = C_2(0, U), \quad g(U) = \frac{\mathcal{A}_2(0, U)}{C_2(0, U)}.$$

Lemma 4.3. *Under the assumptions A2 and A3′, all positive solutions of (4.59) converge to the equilibrium $\bar{E}^* = (P_2^*, H^*, cH^*)$, where $H^* = g(P_2^*)$.*

Proof. Define a function $L : \mathrm{Int}\,\mathbf{R}_+^3 \to \mathbf{R}$, which is associated with (4.59), by

$$
\begin{aligned}
L(U, V, W) \ &= \int_{P_2^*}^{U} \frac{1}{C(s)}\left[C(s) - C(P_2^*)\right] ds \\
&+ \frac{1}{B_2}\left[W - g(P_2^*)\frac{W}{V} - cg(P_2^*)\ln V\right].
\end{aligned}
\qquad (4.60)
$$

Then L is well defined. Note that the derivative of L along (4.59) is

$$\dot{L}(U, V, W) = \left[C(U) - C(P_2^*)\right][g(U) - g(P_2^*)] - \frac{g(P_2^*)}{dB_2 V^2}\left[cV - W\right]^2. \qquad (4.61)$$

The equation (4.61) and assumption A3′ imply that $\dot{L}$ is negative and that the set

$$\Sigma = \left\{(U, V, W) : \ \dot{L}(U, V, W) = 0\right\} = \{U = P_2^*, \ W = cV\}. \qquad (4.62)$$

It is obvious that, by (4.62), the set $\Sigma \setminus \bar{E}^*$ does not contain any globally defined solution of (4.59) (i.e., a solution (4.59) defined for all $t \in \mathbf{R}$). Thus, from Lasalle's Invariance principle it follows that all positive and bounded solutions of (4.59) converge to $\bar{E}^*$ as $t \to \infty$. The proof is completed.

For the proof of the existence of traveling wave solutions of (4.43) the following lemma (its proof can be found in [96]) is needed.

Lemma 4.4. *For each $c > 2\sqrt{\rho(M - d)}$, let E_1^{*U} be the unstable manifold of the equilibrium E_1^*. Then there is a point $p_* \in \mathrm{Int}\,\Omega \cap E_1^{*U}$ such that the solution $\Psi_*(t)$ of (4.49) through the point p_* stays in $\mathrm{Int}\,\Omega$ for all $t \geq 0$.*

Now we are ready to establish the following theorem for the existence of traveling wave solutions.

Theorem 3 *Under the conditions A1, A2, and A3′, for each $c > 2\sqrt{\rho(M - d)}$, the system (4.49) has a nonnegative heteroclinic orbit satisfying the condition (4.50). Hence, the reaction-diffusion system (4.43) has a traveling wave solution connecting the equilibrium points E_1^* and E_2^*.*

Proof. Let $\Psi_*(t)$ be a solution of (4.49) defined in Lemma 4.4. Then $\Psi_*(t) \to E_1^*$ as $t \to -\infty$ because $\Psi_*(t)$ is in the unstable manifold of E_1^*. Moreover, $\Psi_*(t) \in \mathrm{Int}\,\Omega$ for all $t \geq 0$. Hence, $\Psi_*(t)$ is positive and bounded, and its component $U_1(t) \to 0$ as $t \to \infty$ by Lemma 4.2. It follows that its ω-limit set, ω, exists and

$$\omega \subset \{U_2 = 0, \ 1 \leq U_1 \leq 1, \ 0 \leq \gamma_1 V \leq W \leq \gamma_2 V\}.$$

Lemma 4.3 and a result from Corollary 4.3 in [353] imply that ω must be an equilibrium point. By Lemma 4.2, $\omega \neq \{(0,0,0,0)\}$ and $\omega \neq \{(0,1,0,0)\}$. Hence, from Lemma 4.3 it follows that $\omega = (0, P_2^*, H^*, cH^*) = E_2^*$. Consequently, $\Psi_*(t) \to E_2^*$ as $t \to \infty$. The proof is completed.

From Theorem 3 it follows that the reaction-diffusion system (4.43) has traveling wave solutions connecting from $E_1 = (1,0,0)$ to $E_2 = (0, P_2^*, H^*)$. This provides a description of dynamic landscape patterns with which changes in the vegetation composition may occur. In particular, under the assumptions (I) and (II), the results provide information about the conditions under which the invasion of toxic plant species through the space is possible. The necessary conditions needed for the existence of traveling wave solutions include Assumptions A1–A3. It is shown in [96] that Assumption A1 is equivalent to

$$c_{21} < 1 < c_{12}, \quad \frac{\sigma_2}{r_2} < \frac{\sigma_1}{r_1}. \tag{4.63}$$

Assumption A2 implies that the functional response curve $C(P)$ (for plant species 2) has a unique intersection with the line d/B (see Figure 4.3), and Assumption A3 suggests that the P^* component of the unique equilibrium (for plant species 2) is on the right of the point P_h (see Figure 4.5) so that the equilibrium is stable and no periodic solution exists.

4.5 Glossary

2-D TDFRM: 2-dimensional toxin-determined functional response model or 2-dimensional TDFRM.

3-D TDFRM: 3-dimensional toxin-determined functional response model or 3-dimensional TDFRM.

Additive toxins: These are two different toxins that have additive effects. This is in contrast with non-interactive toxins, where each toxin acts independently and neither enhances the negative effects of the other.

Global asymptotic stability: A system is globally asymptotic stable when the domain of stability of an equilibrium point is the whole space surrounding the point. This is stronger than local stability, which means only that the trajectory will return to the equilibrium point following small perturbations.

HT2FR: Acronym for Holling type 2 functional response.

Homoclinic bifurcation: A homoclinic bifurcation occurs when a periodic orbit intersects with a saddle point, joining the saddle point to itself, leading to an orbit (homoclinic orbit) of infinite duration.

Jacobian matrix: In sets of non-linear differential equations, the Jacobian matrix is found as the square matrix of partial derivatives of the right-hand side of the equations with respect to each of the variables and evaluated at an equilibrium point.

Poincaré–Bendixon theorem: Let D be a closed bounded region of the (x, y)-plane, where variables x and y are described by coupled non-linear differential equations in which the right-hand sides are continuously differentiable. Any trajectory remaining in that domain must be a closed orbit, approach a closed orbit, or asymptotically approach an equilibrium point.

Rare plant effect: This effect can occur when a plant is both rare and less defended (e.g., by toxin) from herbivory than other plants in its environment. Herbivores may browse on this plant disproportionately to its frequency. This is an example of toxin-dependent depensation.

Reaction-diffusion systems: These are models of one or more populations in which the populations undergo dynamics as well as diffuse in space.

Resource availability hypothesis: This hypothesis states that plants growing on infertile soils, and thus slow-growing, will invest more heavily in anti-herbivore defenses than those growing on fertile soils, and thus fast-growing.

Saddle point: In mathematics, a saddle point or minimax point is a point in the domain of a function where the slopes (derivatives) of orthogonal function components defining the surface become zero (a stationary point) but are not a local extremum on both axes. The saddle point will always occur at a relative minimum along one axial direction (between peaks) and where the crossing axis is a relative maximum. The name derives from the fact that the prototypical example in two dimensions is a surface that *curves up* in one direction, and *curves down* in a different direction, resembling a riding saddle or a mountain pass between two peaks forming a landform saddle. In terms of contour lines, a saddle point in two dimensions gives rise to a contour graph or trace that appears to intersect itself-such that conceptually it might form a "figure eight" around both peaks; assuming the contour graph is at the very "specific altitude" of the saddle point in three dimensions.

Stable focus: In a two-dimensional dynamic system, a stable focus occurs for an equilibrium point if the Jacobian matrix has complex conjugate eigenvalues with a negative real part. All trajectories starting close to the equilibrium spiral toward the equilibrium

Stable node: If all eigenvalues of the Jacobian matrix are negative, the critical point is a stable node.

TDFRM: Acronym for "toxin-determined functional response model."

Transcritical bifurcation: This is a bifurcation that occurs when a zero solution exchanges stability with a non-zero solution. For example, in the logistic equation $\dot{P} = aP - bP^2$, as a passes from negative values through zero, the solution $P = 0$ changes from stable to unstable and the solution $P = a/b$ changes from unstable to stable.

Traveling wave solution: In population theory, this is a solution of a reaction-diffusion equation in which an initial invasion of a population grows and propagates away from the starting point, approaching a fixed speed.

Unstable focus: In a two-dimensional dynamic system, an unstable focus occurs for an equilibrium point if the Jacobian matrix has complex conjugate eigenvalues with a positive real part. All trajectories starting close to the equilibrium spiral away from the point.

Unstable node: An unstable node in a two-dimensional dynamic system is an equilibrium point that has two positive eigenvalues. All trajectories move away from the point. This is the opposite of a stable node.

Part II

Applications

Chapter 5

Plant Quality and Plant Defenses: Parallels and Differences

5.1 Induced defense in plants: Effects on plant-herbivore interactions

5.1.1 Introduction

Chemical defenses can be classified as either constitutive or inducible. Constitutive chemical defenses are ones that are always present, while induced chemical defenses are produced in response to the presence of or damage from natural enemies, in order to deter the natural enemy. Inducible defenses occur in many taxa in both terrestrial and aquatic food webs [184, 185, 359, 373]. Most induced defenses affect handling times and/or attack rates of consumers and thus their functional responses and per capita consumption rates [175]. For example, in aquatic planktonic systems, herbivore-released infochemicals may induce colony formation in freshwater algae [152, 209, 365, 373] and in marine algae [350], and these aggregations are hard for consumers to handle.

Many terrestrial plants have induced chemical defenses that deter feeding by herbivores. In many cases these chemicals can be induced almost immediately or within hours [22, 132]. The induction of defense can also be delayed; in deciduous trees damage done by herbivores in one year may result in greater chemical resistance the following year [141]. The chemical defenses may be induced only locally, through increased levels of phenolics, peroxidases, etc., near the wound, causing an herbivorous insect to give up eating in that particular place; or the defenses may spread systemically in the plant [28, 173]. Proteinase inhibitors are a common inducible secondary compound. These bind with proteinases (digestive enzymes) of microorganisms and insects. Some plants have been shown to have chemical "stress monitors" that can indirectly trigger production of defensive chemicals.

At the level of the population, inducible defenses constitute an important source of heterogeneity of edibility within a natural population. Because not all prey individuals are exposed to consumers at the same time or to the same degree, and induced defenses are often reversible (for example chemical defenses that decay through time) there can be continuing fluxes both from non-defended to defended subpopulations and the reverse. Defense is induced by the consumers, so the fractions of defended and undefended individuals within prey populations may shift back and forth in response to fluctuating signals of consumer densities. Such a dynamic heterogeneity within the prey of food webs causes a type of food web flexibility that has received little attention in food web theory. Approaches that ignore the dynamics of heterogeneity within food web nodes may miss some of the mechanisms that underlie the observed patterns and processes in real food webs.

For autotrophic prey, which are at the bottom of the food web, inducible defenses are important not only because they affect their own fitness and that of their herbivore consumers, but also because they can affect both the structure and stability of food webs as a whole. The first of these effects is that the presence of induced defense can change the response of the steady-state values of the plant and herbivore populations, and thus higher trophic levels, to changes in the productivity of the plants. As discussed in Chapter 3, one of the important generalizations regarding food webs is well known as the hypothesis of exploitation ecosystems (EEH), which specifies how top-down control will affect trophic levels all the way down to the autotrophs [213, 274, 276]. For bitrophic systems (autotrophs plus herbivores) in stable equilibrium the model predicts that autotroph biomass is exclusively controlled by herbivores and will not respond to nutrient enrichment. In a tritrophic system the equilibrium herbivore biomass is exclusively controlled by carnivores in tritrophic systems and should not respond to increases in primary productivity, although both autotroph and carnivore biomass will increase.

However, both field studies and theoretical work show that the pattern predicted by such models should not be expected in all cases. Even a laboratory study of bitrophic and tritrophic microbial food chains, which was designed to minimize confounding factors such as omnivory or the presence of inedible species, showed that increasing primary productivity led to an increase of the population abundances of all trophic levels [186].

Such discrepancies between EEH predictions and the outcome of both laboratory and field studies requires the identification of ecological factors not present in the [274] model that governs trophic level biomass responses in nature. Heterogeneity within trophic levels is one factor that theory predicts may have important community level consequences [10, 135, 213, 215, 276, 295, 342]. The presence of both edible and inedible autotroph species introduces heterogeneity within that trophic level, and such heterogeneity has been predicted to change biomass responses to enrichment [7, 213, 214].

The second major influence of induced defenses is the effect they have on the local stability of the interacting populations. As discussed in Chapter 2, local stability is the tendency for a system to return to steady-state, following a perturbation. Plant-herbivore, and predator-prey systems in general, can undergo oscillations under certain circumstances. In Chapter 3 we noted the "paradox of nutrient enrichment," in which increasing carrying capacity of the plant population could lead to a Hopf bifurcation and thus limit cycle oscillations of the plant and herbivore populations.

Vos et al. [374] examined the effects of induced defenses on the EEH hypothesis, which we will consider first. They investigated how inducible defenses affect the changes in the distribution of biomass over the trophic levels when the system is enriched. The authors focused on three aspects of inducible defenses that have been shown to be important in a variety of empirical studies. 1) The induction of defenses depends on consumer density (e.g.,

[14, 206, 365]), 2) both undefended prey and prey with induced defenses may be present at a given moment, over a range of consumer densities [152, 209], and 3) defended prey are not invulnerable [175]. Vos et al. [374] incorporated these aspects of inducible defenses into a classical food chain model [204, 274, 316, 317]. They studied this model analytically and parameterized it for well-studied bitrophic and tritrophic systems to investigate the effects of inducible defenses in an ecologically relevant domain and to provide an example where theoretical results will be amenable to empirical testing. They addressed the following question: Does an increase in primary productivity cause gradual biomass increases of adjacent trophic levels, when inducible defenses are incorporated in the classical Oksanen et al. [274] food chain model? Their model analyses answered this affirmatively. Vos et al. [374] were interested in how induced defense influences food webs up to three trophic levels.

In another study Vos et al. [372] showed that inducible defenses may prevent a paradox of enrichment (i.e., enrichment causing the system to become unstable and oscillate) in both bitrophic and tritrophic food chains, although such enrichment-driven instability exists in the cases in which defenses were permanent or absent [372]. It was shown analytically that the absence of a paradox of enrichment depended on differences in handling times and/or conversion efficiencies on defended and undefended prey. Abrams and Walters [9] had earlier analyzed a bitrophic model that differed in two ways from the model below; prey with induced defenses were assumed to be completely invulnerable, and the induction and decay of defenses did not depend on predator densities. That study also showed that a paradox of enrichment may be absent when defenses are inducible.

We are interested here only in the bitrophic chain of autotroph and herbivore, so we analyze two models from the studies of Vos et al. [372, 374] that simplify their model to plants and herbivores (though our second model includes nutrients as well), in which the autotrophs can be in a defended or undefended state.

5.1.2   Model 1

The model used here is similar to that of equations (2a,b,c) in both papers of Vos et al. [372, 374]. It is also similar in form to equations (2.11) of Chapter 2, except that here we assume that there are fluxes between the two plant subpopulations that depend on both the size of the subpopulations and the density of the herbivore;

$$\frac{dP_1}{dt} = P_1\left[r_1\left(1 - \frac{P_1}{K} - \frac{P_2}{K}\right) - \frac{f_1 H}{1 + \sum_{i=1}^{2} f_i h_i P_i} - d_1\right] - P_1 I(H) + P_2 D(H) \qquad (5.1a)$$

$$\frac{dP_2}{dt} = P_2\left[r_2\left(1 - \frac{P_1}{K} - \frac{P_2}{K}\right) - \frac{f_2 H}{1 + \sum_{i=1}^{2} f_i h_i P_i} - d_2\right] + P_1 I(H) - P_2 D(H) \qquad (5.1b)$$

$$\frac{dH}{dt} = H\left[\frac{c_1 f_1 P_1 + c_2 f_2 P_2}{1 + \sum_{i=1}^{2} f_i h_i P_i} - d_3\right], \qquad (5.1c)$$

Here P_1 and P_2 are the undefended and defended autotroph subpopulations, and H is the herbivore; a schematic is shown in Figure 5.1. P_1, P_2, and H are all biomass densities. The parameter K is the carrying capacity for the total plant population $P_{total} = P_1 + P_2$, and r_1 and r_2 are the maximum growth rates. The parameters d_1 and d_2 are the linear coefficients of mortality of the two plant types. Although the evidence concerning the possible tradeoffs in cost for induced defenses is very weak or lacking [11], we may assume here that the growth rate coefficient is less for the defended plants and/or the mortality coefficient is greater. The parameters f_1 and f_2 are rate coefficients of consumption by the herbivore of non-defended and defended plants, respectively, and h_1 and h_2 are the handling times of the two plant types. The parameters c_1 and c_2 are the efficiencies of plant conversion into herbivore biomass. Consumption rates of plants follow Holling type 2 functional response for two prey (two plant species). Defended plants P_2 can impose a decreased capture rate f_2 on herbivores, relative to the higher f_1 of undefended plants P_1. We will assume that $h_1 < h_2$; that is, a particular amount of biomass of the defended plants takes longer for the herbivore to handle.

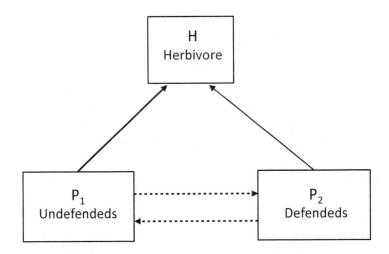

FIGURE 5.1: Schematic diagram of Model 1. The solid lines represent exploitation. The dashed lines represent switching between the undefended and defended subpopulations.

The per unit biomass rate of induction of defenses is I, and the per unit biomass rate of decay of defenses is D, and both are assumed to be functions of the herbivore density. Note that induction is a process that subtracts from the undefended subpopulation of the resource population and adds to the defended subpopulation. Decay of defenses takes away from the defended part of a prey population and adds to the undefended part. By definition, a food chain without defenses is composed of only P_1, while only P_2 exists in a food chain with permanently defended autotrophs. The assumption is made in this model that individuals are able to move from undefended to defended (induction of defenses) or the reverse (decay of defenses). It is unlikely in nature that induction or decay occur at razor-sharp values of herbivore density, because members of the autotroph population will not all sense the same herbivore density. Therefore, it is preferable to model these fluxes as continuous functions of this density. Induction of defenses is assumed minimal at low consumer densities and maximal at high consumer densities. The reverse is true for the decay of defenses. According to this model, fluxes in both directions are usually occurring simultaneously. We assume that the induction and decay rates are sigmoidal in shape and have the specific mathematical forms:

$$I(H) = i\left(1 - \left(1 + \left(\frac{H}{g}\right)^b\right)^{-1}\right) = \text{induction rate of autotroph defenses} \qquad (5.1\text{d})$$

$$D(H) = i\left(1 + \left(\frac{H}{g}\right)^b\right)^{-1} = \text{decay rate of autotroph defenses}, \qquad (5.1\text{e})$$

where i governs the rate of switching between the subpopulations, g is the density of herbivores at which autotroph defense induction reaches half its maximum rate (thus can be called a "half-saturation" constant), b is a shape parameter of the defense induction and decay functions (Figure 5.2). Note that $I(H) + D(H) = i$. The fluxes equations in (5.1) ensure that eventually all autotrophs become undefended when no herbivores are present. When consumers are present, a balance of defense induction and decay will be approached at a rate that depends on parameter i.

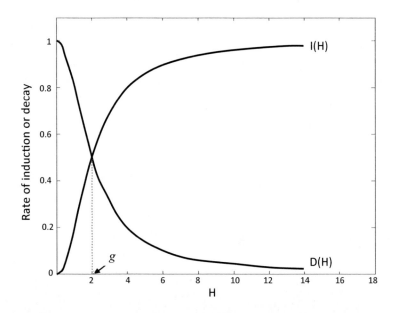

FIGURE 5.2: Rate coefficients for defense induction $I(H)$ and decay $D(H)$ as functions of herbivore density ($g = 5.0$, $b = 2.05$, $i = 1$).

In this system, when all autotrophs are either undefended or defended and cannot change, the variables P and H respond to a change in the carrying capacity, K, according to the EEH. For very low values of K, the herbivores cannot be supported, and plants will be non-defended, with a biomass given by

$$P_1^* = K\left(1 - \frac{d_1}{r_1}\right).$$ (5.2)

As K is increased, a value K_{tc}, called a transcritical point,

$$K_{1c} = \left(\frac{r_1}{r_1 - d_1}\right)\left(\frac{d_3}{f_1(c_1 - d_3 h_1)}\right).$$ (5.3)

will eventually be reached, at which herbivores can be sustained at a level

$$H^* = \frac{1 + f_1 h_1 P_1^*}{f_1}\left[r_1\left(1 - \frac{P_1^*}{K}\right) - d_1\right].$$ (5.4)

K_{tc} can be found by plugging from (5.2) into the right-hand side of

$$\frac{dH}{dt} = H\left[\frac{c_1 f_1 P_1}{1 + f_1 h_1 P_1} - d_3\right],$$ (5.5)

and finding the value of K for which the right-hand side of (5.5) switches from negative to positive. As shown in Chapter 3, once the herbivore is able to invade the initially plant-only system, the top-down effect of the herbivore on the plant holds the plant biomass at a level determined solely by the herbivore parameters;

$$P_1^* = \frac{d_3}{f_3(c_1 - d_3 h_1)} \tag{5.6}$$

(see Figure 5.3). Because H is now non-zero, $I(H)$ is now greater than zero if $i > 0$, so that some fraction of the plant population becomes defended when the herbivore is present. As long as H is small, this flux to the defended subpopulation is small, but as H increases, due to increasing K, a larger fraction of the plant population becomes defended. Because there is a counterflux, $D(H)$, due to decay of the induced defense, an equilibrium can be reached in which the fluxes in the two directions balance.

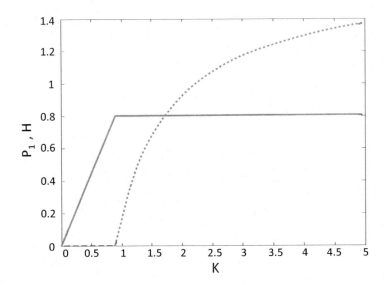

FIGURE 5.3: Behavior of equilibrium values of undefended plants, P_1^* (solid line), and herbivores, H^* (dashed line) as a function of carrying capacity K. Initially P_1^* grows according to equation (5.2). Then, when the transcritical point, K_{tc}, is reached, H^* increases according to equation (5.4) and P_1^* is fixed at equation (5.6). Parameter values are $r_1 = 1.42$, $d_1 = 0.145$, $h_1 = 0.5$, $f_1 = 0.77$, $c_1 = 0.36$, $d_3 = 0.17$.

If $i > 0$, it can be shown that the result of EEH is changed when the prey, plants in this case, have inducible defenses. Contrary to the EEH, when the carrying capacity of the plant, K, increases, both total plant biomass, $P_{total}^* = P_1^* + P_2^*$ and H^*, increase. To show this, Vos et al. [374] let $x = P_2/(P_1 + P_2)$. Then one can solve for P_{total}^* by setting the right-hand side of equation (5.1c) to zero:

$$P_{total}^*(x) = \frac{d_3}{c_1 f_1 + (c_1 f_2 - c_1 f_1)x - d_3 f_1 h_1 - d_3(f_2 h_2 - f_1 h_1)x}, \tag{5.7a}$$

where, for simplicity, we have set $c_2 = c_1$. If we also make the assumption that $f_1 = f_2$, and $h_1 < h_2$, meaning that the acquisition rates of the two plant types are the same but the handling time for the defended plant biomass is greater, then equation (5.7a) becomes

$$P_{total}^*(x) = \frac{d_3}{c_1 f_1 - d_3 f_1 h_1 - d_3 f_1 (h_2 - h_1)x}. \tag{5.7b}$$

The ratio x will increase with an increase in herbivore biomass, H^*, as defended plants are favored. It can be seen that, because $h_2 > h_1$, any increase in the fraction of defended plants, x, will lead to a decrease in the denominator of (5.7b) and hence an increase in P^*_{total}.

Thus when both undefended and defended autotrophs are present and have the ability to switch between the defended and undefended states, the total autotroph biomass increases gradually over that range of values of K. This is verified in Figure 5.4, for parameter values shown in the caption. The carrying capacity, K, is gradually increased during a simulation and both P_1 and P_2 increase through time. This shows that inducible defenses have the effect of modifying the top-down control of the herbivore such that the herbivore does not control the total biomass of the autotrophs.

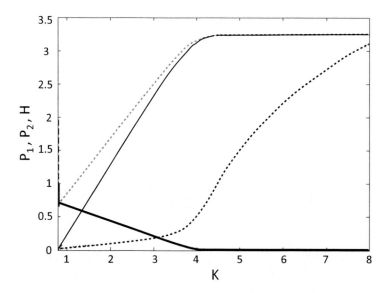

FIGURE 5.4: Response of undefended, P^*_1 (thick solid line), and defended P^*_2 (thin solid line) plants at equilibrium when carrying capacity, K, is increased, for the full set of equations (5.1a–e). The dotted line is the herbivore H^*. Note that the sum, $P^*_1 + P^*_2$ (dashed line) increases with increasing K. Parameter values are $r_1 = 1.42, r_2 = 1.42, d_1 = 0.145, d_2 = 0.18, h_1 = 0.5, h_2 = 1.04, f_1 = 1.5, f_2 = 0.77, c_1 = 0.36, c_2 = 0.36, d_3 = 0.25, g = 0.06, i = 0.1$.

It is also possible to show that inducible defense of plants increases the size of carrying capacity, K, needed to create a Hopf bifurcation, leading to limit cycle oscillations. First, when only the undefended plants are present, the value of K at which the Hopf bifurcation occurs is

$$K_{Hopf} = \frac{r_1}{r_1 - d_1}\left[\frac{2d_3}{f_1(c_1 - d_3 h_1)} + \frac{1}{f_1 h_1}\right], \qquad (5.8)$$

and a similar formula occurs if all the plants were defended, with substitution of r_2, d_2, f_2, h_2, and c_2. To demonstrate the effects of inducible defenses, bifurcation diagrams are shown for both K_{tc} and K_{Hopf} for the cases of all non-defended plants (Figure 5.5a), all defended plants (Figure 5.5b), and the inducible case (Figure 5.5c), which has to be determined numerically, for the parameters K and d_3. Note that for all values of d_3, the value of K_{Hopf} is much higher for the inducible case than each of the two pure strategies. The herbivore cannot exist for $c_1 < d_3 h_1$.

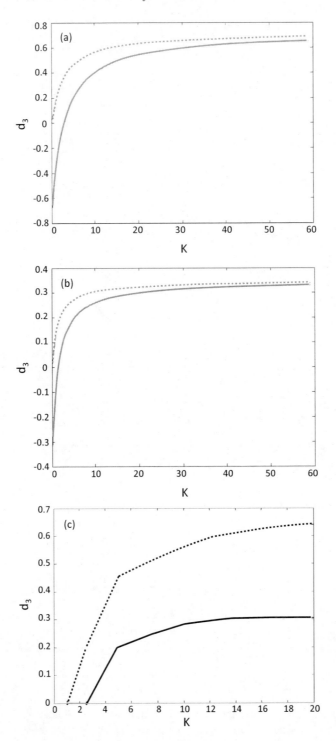

FIGURE 5.5: (a) (K, d_3)-parameter plane diagram for transcritical point, K_{tc} (dashed line) and Hopf bifurcation point, K_{Hopf} (solid line) for interaction of undefended plant, P_1, and herbivore. Parameter values are $r_1 = 1.42, d_1 = 0.145, h_1 = 0.5, f_1 = 0.77, c_1 = 0.36, d_3 = 0.17$. (b) For interaction of defended plant, P_2, and herbivore, where $h_2 = 1.04, d_2 = 0.18$ and other parameters are the same. (c) For interaction of both undefended plant, P_1, defended plants, P_2, and herbivores, where now $i = 0.1, g = 0.06$.

5.1.3 Model 2

An extension of the model is obtained by including a limiting nutrient of the autotroph, which also leads to more stable behavior of the model. The revised model can be written

$$\frac{dN}{dt} = Q(N_0 - N) - \frac{(v_1 P_1 + v_2 P_2)N}{k + N} \tag{5.9a}$$

$$\frac{dP_1}{dt} = \frac{v_1 P_1 N}{k + N} - \frac{f_1 H P_1}{1 + f_1 h_1 P_1 + f_2 h_2 P_2} - d_1 P_1 + D(H)P_2 - I(H)P_2 \tag{5.9b}$$

$$\frac{dP_2}{dt} = \frac{v_2 P_2 N}{k + N} - \frac{f_2 H P_2}{1 + f_1 h_1 P_1 + f_2 h_2 P_2} - d_2 P_2 - D(H)P_2 + I(H)P_2 \tag{5.9c}$$

$$\frac{dH}{dt} = \frac{c_1 f_1 H P_1 + c_2 f_2 H P_2}{1 + f_1 h_1 P_1 + f_2 h_2 P_2} - d_3 H, \tag{5.9d}$$

where N is the limiting nutrient concentration, P_1 and P_2 are the undefended and defended autotroph subpopulations, and H is the herbivore (Figure 5.6). P_1, P_2, and H are all expressed in terms of the limiting nutrient of the autotroph, so all variables have units of nutrient per unit area in this case. The parameter Q represents the flux of water into the system and N_0 is the input concentration of the water. Nutrient recycling is not considered; that is, all losses not consumed by higher levels leave the system. The parameters, v_1 and v_2, are rates of nutrient uptake by the non-defended and defended plants, respectively, with the assumptions that $v_1 > v_2$, and k is a half-saturation coefficient for nutrient uptake, the same for both defended and undefended plants. If N_0 is increased from zero, then initially, the equilibrium nutrient level, $N^* = N_0$ will be too small to support plants. However, when N^* reaches a concentration such that

$$\frac{v_1}{k + N^*} - d_1 > 0;$$

that is, it is a transcritical point. We have also assumed that

$$\frac{v_1}{k + N^*} - d_1 > \frac{v_2}{k + N^*} - d_2,$$

so that the undefended autotroph is the invader. When N_0 increases further a second transcritical point is reached, at which

$$\frac{c_1 f_1 P_1}{1 + f_1 h_1 P_1 + f_2 h_2 P_2} - d_3 > 0,$$

where the herbivore can invade the system. We continue to assume that $i = 0$, so there is no switching between undefendeds and defendeds, and only undefendeds are present. Then the variables N, P, and H respond to a change in nutrient input, N_0, according to the EEH.

$$P_1^* = \frac{d_3}{c_1 f_1 - f_1 h_1 d_3} \tag{5.10a}$$

$$N^* = -\frac{1}{2}\left(\frac{v_1 P_1^*}{Q} + k_1 - N_0\right) + \frac{1}{2}\left[\left(\frac{v_1 P_1^*}{Q} + k_1 - N_0\right)^2 + 4k_1 N_0\right]^{1/2} \tag{5.10b}$$

$$H^* = \left(\frac{v_1 N^*}{k + N^*} - d_1\right)(1 + f_1 h_1 P_1^*). \tag{5.10c}$$

Note that P_1^* does not change with increasing N_0, as it is held by the top-down effect of the herbivore. However, both N^* and H^* increase with N_0.

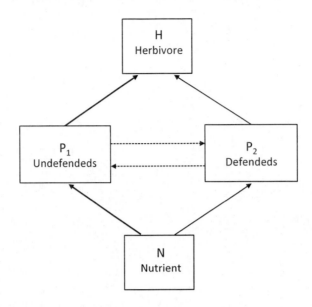

FIGURE 5.6: Schematic diagram of Model 2. The solid lines represent exploitation. The dashed lines represent switching between the undefended and defended subpopulations.

For the time being, we continue to assume there is no switching between subpopulations, because, in general there would be a net flux from one population to the other at equilibrium and the equations would be difficult to solve analytically. For the purpose of allowing an analytic solution, we still assume that $i = 0$. Induction of defense can still occur, but only for new individuals to the population. Further increase in N_0 leads to a third transcritical point, the criterion for defendeds to enter the system;

$$\frac{v_2 N^*}{k_1 + N^*} - \frac{f_2 H^*}{1 + f_1 h_1 P_1^*} - d_2 > 0.$$

Now, both undefended and defended autotrophs are present, although there is assumed to be no switching between the two types. The steady-state values of the system can still be solved for in this case, and, when all variables are non-zero, they are,

$$N^* = \frac{(f_2 d_1 - f_1 d_2) k_1}{f_2 v_1 - f_1 v_2 + f_1 d_2 - f_2 d_1} \tag{5.11a}$$

$$P_1^* = \frac{v_2 d_3 - A_2 A_3}{v_2 A_1 - v_1 A_2} \tag{5.11b}$$

$$P_2^* = \frac{A_3 - v_1 P_1^*}{v_2} \tag{5.11c}$$

$$H^* = \left(\frac{v_1 N^*}{k_1 + N^*} - d_1\right)\left(\frac{1 + \sum_{i=1}^{2} f_i h_i P_i^*}{f_1}\right) = \left(\frac{v_1 N^*}{k_1 + N^*}\right)\left(1 + \frac{f_2 h_2 d_3}{A_2}\right)\frac{1}{f_1}, \tag{5.11d}$$

where

$$A_1 = c_1 f_1 - d_3 f_1 h_1$$

$$A_2 = c_2 f_2 - d_3 f_2 h_2$$

$$A_3 = Q\left(\frac{N_0 - N^*}{N^*}\right)(k_1 + N^*)$$

$$A_4 = v_1 - v_2 A_1/A_2,$$

and the parameters are such that P_1^* and P_2^* coexist. It can be seen that now N^* is independent of N_0 in this range of nutrient input, so it does not increase as N_0 increases. However, P_1 and P_2 both change with N_0 and the sum

$$P_{total}^* = P_1^* + P_2^* = P_1^*\left(1 - \frac{v_1}{v_2}\right) + \frac{A_3}{v_2} \tag{5.12}$$

increases as N_0 increases. This is the case for reasonable values (Table 5.1), for which $A_2 = c_2 f_2 - d_3 f_2 h_2 > 0$, $v_2 A_1 - v_1 A_2 > 0$, and $v_2 d_3 - A_2 A_3 > 0$. Total autotroph nutrient increases with increases in N_0, because A_3 increases with increases with N_0 and, although P^* decreases with N_0, it is assumed that $v_1 > v_2$, that is, the maximum growth rate of the undefended is higher than that of the defended autotrophs. Therefore, both terms on the right-hand side of (5.12) increase with N_0.

Using equations (5.10) and (5.11), plus $N^* = N_0$ when

$$\frac{v_1}{k + N^*} - d_1 < 0,$$

we can plot the values of the parameters as a function of increasing N_0, in this case from $N_0 = 0$ to 2.4 (Figure 5.7). The changes in the system structure for increasing N_0 can be seen. Initially N^* increases identically with increasing N_0. When P_1^* invades the system, it holds N^* constant. Next H^* is able to invade, which holds P_1^* constant, so that N^* is able to resume increasing with N_0, though with a lower slope. Slightly past $N_0 = 0.75$, P_2^* is able to invade. This occurs because, even though P_1^* has a higher growth rate than P_2^*, the herbivore has a stronger negative effect on the undefended prey, P_1^* (see Table 5.1). Therefore, P_2^* increases from zero, and P_1^* starts to decline. What is remarkable then is that N^* is again held constant over the following interval. The reason is that, while the herbivore is able to hold a linear combination of P_1^* and P_2^* constant, i.e.,

$$\frac{c_1 f_1 P_1^* + c_2 f_2 P_2^*}{1 + f_1 h_1 P_1^* + f_2 h_2 P_2^*} = d_3,$$

it cannot hold the individual autotroph species constant, and the combination of P_1^* and P_2^* is able to hold N^* constant. The next transition occurs when P_1^* goes to extinction. At that point, the herbivore holds P_2^* constant, and N^* again resumes increasing, while the rate of increase of H^* jumps, due to the resumed increase of N^*.

This analysis of the equilibria does not tell us about instabilities and it ignores the existence of switching behavior. To consider the possibility of instability, it is easier to simply perform simulations over the range of values of N_0 than deriving the complicated stability conditions. Let us look over the range of values of the variables, with N_0 starting from 0.7, but initially continue to ignore switching behavior; that is assume that $i = 0$, so there are no transitions between defendeds and undefendeds, though both subpopulations can exist over a range of values of N_0. The bifurcation diagram is found numerically by simulation with gradually increasing N_0 (Figure 5.8a). At first, only N^* increases, and the other variables are zero. Then P_1^* can invade and increase, with N^* held to a constant value.

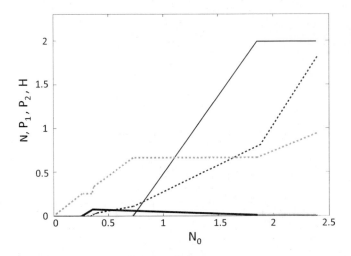

FIGURE 5.7: Changes in N^*, P_1^*, P_2^*, and H^* as N_0 is increased, in Model 2. Initially, only N^* (dashed line) increases. Then, as the transcritical point is reached, P_1^* (thick solid line). Further increase allows H^* to invade. Because H limits P_1^*, P_2^* (thin solid line) can invade, at which point N^* is held constant. P_1^* is driven to extinction, at which point P_2^* is held constant and N^* increases again. Parameter values are given in Table 5.1.

The herbivore then invades for a slightly higher value of N_0 and, as N_0 is increased further, P_1^* is held constant and N^* increases. At a higher value of N_0, P_2^* is able to invade and increase, with P_1^* now decreasing and N^* constant. As N_0 is increased still further, P_1^* goes to extinction. Now there are only defended plants, and the system is again a linear trophic chain. Therefore, N^* again increases with increasing N_0, and H^* increases at a higher rate, and P_2^* is held constant. This pattern continues until N_0 slightly exceeds 3.1, at which point the system becomes unstable and N^*, H^*, and P_2^* all oscillate. Therefore, the simulations duplicate the results from the analytic values of the equilibria, with the addition of showing a paradox of enrichment Hopf bifurcation, at a sufficiently large value of N_0. However, the Hopf bifurcation is shifted to higher values of N_0 than if there were no defendeds. This can be seen by performing the simulation without the defendeds. In this case the instability occurs for a much lower value of N_0 (Figure 5.8b). The fact that $f_1 > f_2$ and $h_2 > h_1$ causes the interaction of herbivores with defendeds to be stable for higher values of N_0 than is the interaction of herbivores with undefendeds. This agrees with the results of Edelstein-Keshet and Rausher [93] that induced defenses tend to be stabilizing. It can be speculated that if there was an additional level of induced defense that was even more protected from the herbivore, the Hopf birfurcation could be pushed to a still higher concentration of N_0.

We want to extend these results to the case where switching can occur. Suppose first that the switching takes the form of simple diffusion, so that $I(H) = D(H) = 0.05$, that is, the coefficients of switching are constant and equal for each direction. In this case, the instability occurs at a lower value of N_0 than for the case in which there is no switching, slightly above $N_0 = 1.7$ (Figure 5.8c, compare with Figure 5.8a). The reason for this may be that the purely diffusive switching reduces the relative numerical dominance of the defendeds over undefendeds, when both exist, making the system more vulnerable to instability.

We next examine the effect of allowing $I(H)$ and $D(H)$ to assume the forms shown in equations (5.1d,e), so that the rate of switching is determined by the herbivore density H^*. First note that a key idea in spatial ecology is that animals will attempt to distribute themselves on a landscape in such a way that each maximizes its individual fitness. This can be said to happen when no individual can gain fitness by moving from its present position to any other. The distribution across patches is called the ideal free distribution (IFD) [113]. According to the IFD, if other factors besides carrying capacity of a patch, such as predator density, can be ignored, and animals are free to move, and move in a way to maximize fitness, they will continue to move until they cannot do any better in terms of fitness, which is the point at which the subpopulation on each patch matches its carrying capacity. The IFD can apply not just to spatial distributions, but distributions of attributes such as defense from predators. When applied to the population of defended and undefended plants, it means that no individual, whether defended or undefended, would increase its fitness by switching. Switching between the states could be occurring, but it would be perfectly balanced. This will occur when the defense induction rate is balanced by the decay rate of defenses, or, at equilibrium $I(H^*)P_1^* = D(H^*)P_2^*$. Just setting g to a particular value, and setting $I(H^*) = D(H^*)$ will not achieve that. For example, making this assumption and setting $g = 1.5$ in this case, leads to the results shown in Figure 5.8d. The Hopf bifurcation still occurs at a concentration N_0 less than the case of no switching. The reason is the same as in the preceding case, where $I(H) = D(H) = 0.05$; that is, the switching lowers the dominance of the defended, making the system more vulnerable to instability.

Let's examine the parameter g more closely. This parameter is a behavioral parameter of the autotrophs governing the likelihood of switching at a given herbivore density H. For values of herbivore density, $H > g$, the net flow from undefended autotrophs to defended autotrophs is positive, whereas for $H < g$, the net flow is negative. The value of g in the simulations shown in Figure 5.8d was chosen arbitrarily as $g = 1.5$. To obtain $I(H^*)P_1^* = D(H^*)P_2^*$ we need to make g a function of those three equilibrium values; that is,

$$g_{\text{IFD}} = H^* \Big(\frac{P_1^*}{P_2^*}\Big)^{1/b_1}, \tag{5.13}$$

which can be found by solving $I(H^*)P_1^* = D(H^*)P_2^*$ for g. If this value of $g = g_{\text{IFD}}$ is plugged into $I(H^*)$ and $D(H^*)$ and the simulations performed again, the result is shown in Figure 5.8e. This is identical to Figure 5.8a, in which $I(H^*) = D(H^*) = 0$ because setting g to g_{IFD} guarantees that there is no net movement in either direction between undefendeds and defendeds, which is the same as if there is no movement at all. We plot both g_{IFD} and as functions of N_0 (Figure 5.8f). g_{IFD} is infinitely large when $P_2^* = 0$, but declines rapidly as soon as P_2^* invades the system and then goes to zero when P_1^* goes to extinction.

Using $g = g_{\text{IFD}}$ in the switching functions in the model produces exactly the same equilibrium values of the variables as does setting $I(H) = D(H) = 0$ (compare Figures 5.8a and 5.8e). What, then is the difference between models having the two different assumptions? The difference is in the transient behavior following perturbation of the variables away from their equilibrium values, as they return toward the equilibrium. Let's consider a particular value of $N_0 = 1.1$. First, perform the simulation with $I(H) = D(H) = 0$. Note that oscillations occur that dampen over about 500 time units (Figure 5.9a), to the equilibrium values $P_1^* = 0.0428$, $P_2^* = 0.6652$, and $H^* = 0.3345$ for that value of N_0. However, if these values are plugged into $g = g_{\text{IFD}} = 1.32$, and g_{IFD} into $I(H^*)$ and $D(H^*)$ along with a relatively high value of i, such as $i = 0.5$, the transient behavior drastically changes, such that the oscillations are completely eliminated (Figure 5.9b). This is because the switching behavior steers the system toward its equilibrium.

The value of $g = g_{\text{IFD}}$ leads to an IFD, such that the fluxes from undefendeds to defendeds is equal to the flux in the opposite direction at equilibrium. It has been found that under fairly general conditions IFDs are also evolutionarily stable strategies (ESSs); that is, strategies, which, if adopted by a population, cannot be invaded by a population with an alternative strategy, if that population is initially rare [76, 52]. In particular, we ask if

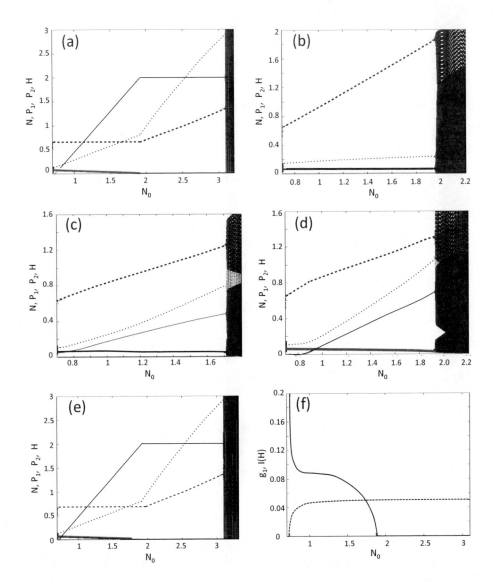

FIGURE 5.8: (a) Changes in N^* (dashed line), P_1^* (thick solid line), P_2^* (thin solid line) and H^* (dotted line) as N_0 is increases, in Model 2. Parameter values are given in Table 5.1, with $i = 0$, so there is no switching. This situation is similar to that of Figure 5.7, but N_0 starts at 0.7. At slightly above N_0 a Hopf bifurcation is reached and the system becomes unstable. (b) Only N^*, P_1^*, and H^* are included in the model. As N_0 is increased, a Hopf bifurcation is reached at about $N_0 = 1.95$. (c) Similar to (a) but switching is allowed in the form of diffusion, with $I_1(H) = D_1(H) = 0.05$. As N_0 is increased, a Hopf bifurcation is reached slightly above $N_0 = 1.75$. (d) Similar to (c) but the value of g_1 is fixed at 1.5 and $i = 0.05$. As N_0 is increased, a Hopf bifurcation is reached slightly below $N_0 = 1.85$. (e) The value of g is now allowed to be such that the switching between undefendeds and defendeds is the same; that is $I(H^*)P_1^* = D(H^*)P_2^*$ are equal, which occurs when $g_{\mathrm{IFD}} = H^*(P_1^*/P_2^*)^{1/b_1}$, and $i = 0.05$. (f) The values taken on by g (solid) and $I(H)$ (dashed) as N_0 is increased through time, while $D(H)$ goes to zero.

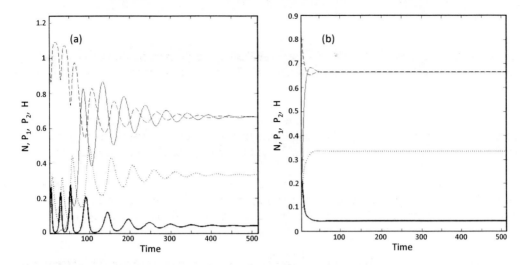

FIGURE 5.9: (a). Transient behavior of model described in Figure 5.8a after a perturbation away from the equilibrium point, where N^* (dashed line), P_1^* (thick solid line), P_2^* (thin solid line) and H^* (dotted line), and $N_0 = 1.1$. Parameter values are given in Table 5.1, with $i = 0$, so there is no switching. (b) Transient behavior of model described in Figure 5.8e after a perturbation away from the equilibrium point, where N^* (dashed line), P_1^* (thick solid line), P_2^* (thin solid line) and H^* (dotted line), and $N_0 = 1.1$. The value of g is now allowed to be such that the switching between undefendeds and defendeds is the same; that is $I(H^*)P_1^* = D(H^*)P_2^*$ are equal, which occurs when $g_{\text{IFD}} = H^*(P_1^*/P_2^*)^{1/b_1}$, and and $i = 0.05$.

there is an ESS such that a plant population with a particular value of g cannot be invaded by any plant population of the same species; that is, for which all parameter values are identical except for a different value of g.

We do this by performing simulations with the set of equations (5.9a,b,c,d) and (5.1d,e), which is amended in order to allow the existence of both a resident and an invading genotype. The resident genotype has the parameter value g_{IFD}, whereas the invader has some other genotype g value. Using a different set of parameters than those used in Figures 5.7–5.9 (see the second parameter value set in Table 5.1), for which $g_{\text{IFD}} = 1.87$, we simulated a number of possible combinations of resident and invader strategies. The results are plotted on a pairwise invasibility plot [122] shown in Figure 5.10, starting with the equilibrium values of the resident strategies and small initial populations of the invaders. The resident's values of $g = g_r$ are shown along the x-axis and the invader's values of $g = g_i$ are along the y-axis. Plus signs show the combinations for which the invaders were successful and minus signs show the combinations for which the residents were successful. Values of "c" indicate apparent coexistence of the two populations. The vertical straight line covers values that are all negative except at the point 1.87, where the resident is competing against an invader identical to itself. The fact that the vertical line drawn through the strategy has only negative signs above and below $g_{\text{IFD}} = 1.87$ means that value is an ESS. Note further that if the resident has values of g_r that are less than $g_{\text{IFD}} = 1.87$, and the invader has values of g_i that are greater than the resident's, the invader always wins; that is, is able to invade. Also, if the resident has values of g_r that are greater than 1.87, and the invader has values of g_i that are greater than the resident's the invader always loses. These imply that the ESS is also a convergence stable strategy (CSS). If the singular strategy is convergence and evolutionarily stable, then the ESS is reached through progressive evolutionary steps, whereby phenotypes closer to the ESS invade and replace those farther from it. This means that $g_{\text{IFD}} = 1.87$ is an ESS-CSS, since natural selection will always cause g to tend toward $g = 1.87$ from any nearby value.

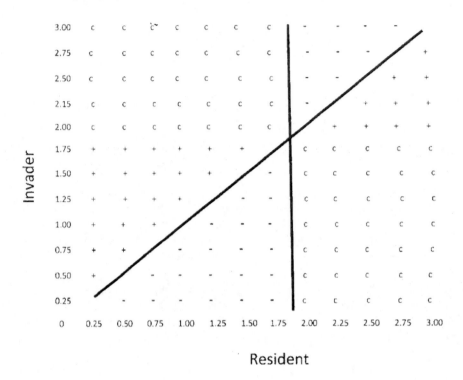

FIGURE 5.10: Invasion plot showing that the value of g, $g_{\text{IFD}} = H^*(P_1^*/P_2^*)^{1/b_1}$, is an evolutionarily stable strategy (ESS) and that it is also convergent. Here the minus signs indicate the pairs of values for which the invader cannot invade and the plus signs indicate pairs of values for which the invaders can successfully invade. The "c" signs represent apparent coexistence. The parameter values are given in Table 5.1. The equilibrium values for that case are $P_1^* = 0.0428$, $P_2^* = 0.6652$, and $H^* = 0.3345$, such that $g_{\text{IFD}} = 1.87$. A population with that value can be shown to resist invasion from any population with a different value of g but the same values of the other parameters.

To summarize this section, we have shown the effect of induced defenses on key features of food chains. Much of the theory regarding interactions between species populations has been built on the basis of models in which the populations are internally homogeneous. Phenomena such as the top-down cascades of the hypothesis of exploitation ecosystems (EEH) and the instability stimulated by increasing carrying capacity at the base of the food web, or the "paradox of enrichment," emerged from these simple models. Using two different models, we have examined one type of heterogeneity within plant populations, that of variation in the level of defense from herbivory. Two phenotypes were assumed, one of which was more defended than the other, though at a cost of tradeoffs such as a lower growth rate. In both models, earlier results of the EEH, in which a homogeneous plant population was assumed, were changed. Rather than the plant population being held to constant level with increasing K (Model 1) or increasing N_0 (Model 2), the total plant population increased with these increases when both phenotypes were present. This occurred because, although

TABLE 5.1: Parameter values in Model 2 used in competition between plant populations with different values of g.

Symbol	value set 1 (used in Figures 5.7–5.9)	value set 2 (used in Figure 5.10)
Q	1	3.0
N_0	variable	10
k_1	1.0	1.0
v_1	2.0	1.1
v_2	1.5	0.9
f_1	4.0	0.1
f_2	1.0	0.0125
h_1	0.1	0.5
h_2	3.5	0.5
d_1	0.4	0.4
d_2	0.5	0.4
d_3	0.1	0.25
c_1	0.6	0.5
c_2	0.6	0.5
i	variable	from 0 to 1
b	2.05	2.05
g	variable	variable

the undefended subpopulation declined as the herbivore population increased, the defended subpopulation rose at a rate that more than compensated for the decline.

These effects of two subpopulations occurred both when there was no switching between the two subpopulations and when there were balanced rates of switching at equilibrium, equivalent to an Ideal free distribution (IFD). However, diffusive switching, that is, switching in which the coefficient of switching was the same in each direction, actually slightly reduced the levels of N_0 at which the Hopf bifurcation instability was reached. For both no switching and IFD switching the behavior of the equilibria with N_0 (and K also, although not shown) was identical. However, the occurrence of switching could drastically reduce the transient oscillations following a perturbation.

The switching function used in the models depended on herbivore density, so that switching increased with H. The strength of the effect of H on switching was governed by a parameter g. When the parameter g was tuned to produce an IFD at equilibrium, that is, when g was given by g_{IFD} in equation (5.13), this was shown to be an ESS as well, that could not be invaded by other strategies. So the ability to switch strategies is not only important to the dynamics of the population, but the rate is subject to natural selection as well.

5.2 Plant defense allocation strategies

We have described models of the effects of plant chemical defenses on plant-herbivore interactions from the point of view of the choices made by the foraging herbivore, given one or more plant populations with defenses. Now we change perspective to describing models that assume there is herbivory and attempt to predict how plants will allocate resources to either deter herbivory or compensate for its effects. We will be giving a brief overview of plant allocation of resources in general.

5.2.1   Allocation of resources by plants

Woody plants allocate their acquired resources, energy (or carbon) and nutrients, to meet their several essential functions. In order to grow and reproduce, plants allocate resources to acquire solar radiation by growing leaf biomass and to acquire nutrients by growing fine root biomass. Investment in stem wood, beyond what is needed structurally by the foliage and roots, is also employed to obtain a competitive advantage through height for capturing light (e.g., [127, 196, 197]). Plants exist within the major constraints that only a certain amount of solar radiation is available per unit area of land, and only a fraction of that energy can potentially be captured by photosynthesis. Therefore, the amount of carbon to allocate is limited, and allocation of carbon to one function necessarily decreases the carbon that could be allocated to other functions.

Ability to adjust carbon allocation in response to environmental conditions is widely documented [114]. The tradeoff in investment between root and shoot (foliage) is well known and has been the subject of many studies and models (e.g., [149, 166, 169, 170, 355, 386]). Nutrient deficiency can stimulate biomass allocation to root biomass and morphology [149], whereas increased resources and stand age have been shown to result in increased partitioning to aboveground wood production and decreased to belowground biomass (see review by [222]). However, shading of leaves by competitors can lead to higher leaf area per unit biomass and higher allocation of aboveground growth to leaves in shade than in the sun in deciduous saplings [195]. Investment in additional stem wood, beyond what is needed structurally by the foliage and roots, to obtain a height advantage in capturing light, is another possible destination of carbon allocation [127]. Empirical data also indicate that other factors should be taken into consideration, such as successional status [128, 222] and tree ontogeny [245].

Models have been developed to advance the scientific understanding of the strategies of tree resource allocation with respect to growth and interactions with their environment. Le Roux et al. [212] identify four basic types of allocation models.

The first of these, empirical models, can be completely based on statistical relationships or can be mechanistic, "based on the physiological and physical processes that underlie the way the system responds to stimuli" [210]. Models such as ECOPHYS [302], applied for example to juvenile poplars, have been derived as research tools to understand the tree's developmental processes. Such models are based on empirical studies under controlled conditions in growth chambers, in which allocation fractions of photosynthate are measured along with other processes. Although the models incorporate some modulation of allocation coefficients by external conditions, their application tends to be limited to a relatively narrow range of conditions.

A second, widely represented type in the literature, is the functional equilibrium model, which assumes that there must be a long-term balance between the acquisition of carbon by foliage and nutrients by roots with the use of these for growth. It is assumed that the relative allocations of carbon and nutrients serve some optimization goal such as the maximization of carbon production. These are defined as teleonomic models by Thornley and Johnson [356]. These models can show how a tree should optimally behave in response to environmental conditions. For example, a model of carbon and nitrogen allocation in a forest stand in steady state [238] shows that increasing nitrogen availability to the roots should optimally lead to a decrease in carbon allocation to fine roots, as might be expected, but to increased allocation to wood rather than to foliage, which remains stable in the model. A model of dry matter portioning applied to Douglas fir and beech [42], which incorporated the effects of site conditions, dominance position, and thinning, showed that over time for trees in a stand, there should be a gradual increase in branches and a decrease in fine roots and foliage in response to thinning. A model for carbon and nitrogen dynamics in a growing even-aged coniferous forest [194] showed that optimal fine root allocation for a competing individual was less than 5% of total production for adequate N, but rose to 30% as nitrogen became more limiting with growth of the stand. This shift to roots had the benefit of decreasing the losses of nutrients from the stand by leaching, although it slowed the production of wood. The model DESPOT [48] simulated allocation of carbon allocation with respect to nitrogen, water and light conditions in a growing, monospecific, self-thinning stand with the aim of determining the allocation strategy that maximized carbon gain in a growing stand. It is more general than many other models in including in its cost-benefit structure the process models for the acquisition of light, nitrogen, and water, and using few allometric constraints. The model predicts the stand should reach a determinate height, but that aboveground primary productivity and leaf area index slowly decline in the mature stand.

The third type, called transport-resistance models, was proposed by Thornley [354] for the acquisition of carbon and nitrogen and use by the plant. The models assume that flows of resources are driven by concentration gradients from their source in shoots and roots to their incorporation in tissue. The resistances along these gradients act to regulate the flows, and growth then acts as a sink for these resources. No goals, such as optimization of any quantity, are assumed. Balanced exponential growth results and from the dynamics, and the variables, such as C, N, P concentrations and root-shoot fractions reach constant values. This approach has been used by Ingestad and Agren [166], Dewar [85], and Luan et al. [231], among others, for modeling individual trees or forest stands. For example, in a root-shoot model of Dewar [85] transport of water, labile nitrogen, and labile carbon is simulated in the xylem and phloem. Transport of a given substrate, say labile carbon, between the shoot and root, depends on the gradient of substrate concentration, soluble carbon in shoot to soluble carbon in root, and the resistance to that flow. The model provides a basis for interpreting root-shoot responses to deficits in different substrates. One advantage of transport-resistance models is that ontogenetic trends emerge automatically, as resistance between sources and sinks increases with age and size of plants.

The fourth kind is called a hierarchical approach by Le Roux et al. [212], in that there is a hierarchy of priorities for meeting demands of various internal components of the plant. A model of this type is that of Wermelinger et al. [379] for grapevines. Pools of carbonates and nitrogeneous compounds are assumed, which are augmented daily by carbon assimilation and nitrogen uptake, plus mobilization of reserves. These are distributed by a priority scheme. Meeting maintenance costs is the first priority for the allocation of carbonates. Reproduction is second priority and vegetative growth third priority after blooming, but have equal priority before blooming. Last priority is replenishment of reserves. The demands on nitrogen compounds are proportional to the demands on carbon. In other models using the hierarchical approach, priorities may also depend on the proximity of a component to the source (e.g., [134, 378]). In the latter model, TREGRO, aboveground components have first access to new photosynthate. Below, we describe in detail the model PLATHO [120] that uses the hierarchical scheme, but also includes allocation to plant defense.

Le Roux et al. [212] discuss the advantages and disadvantages of the approaches.

5.2.2 Allocation when anti-herbivore defenses are included

Allocation of resources to anti-herbivore defense was not included in the brief review of models above. Our focus now is the tradeoff between carbon invested in biomass growth (foliage, roots, and structural wood) and defense; in particular, investment in secondary metabolism for producing carbon-based defense compounds.

In addition to the allocation of resources to these essential functions of maintenance, growth, and reproduction, resources are usually allocated to chemical anti-herbivore defense (e.g., [70, 151, 247]), or other types of defense, such as spines [136] or thickening of leaf surfaces, because herbivory can significantly reduce growth and significantly increase mortality, thereby reducing plant fitness.

However, as Craine [73] noted, despite extensive research on chemically mediated plant-herbivore interactions over the past 25 years, an improved synthesis of plant defenses as a part of plant resource allocation strategies is still needed. Although such a synthesis will ultimately depend upon on empirical examination of patterns of chemical defense over a broad variety of environments, we believe that the efficiency of this empirical research can be greatly increased by appropriate use of modeling. In a review Herms and Mattson [151] argue that the optimization of physiological tradeoffs between growth and defense is a central dilemma of plants; that is, increased allocation of resources to growth and reproduction is at the expense of defense, and reduced defense can result in such severe herbivory that fitness is significantly reduced.

Several hypotheses, or more accurately, sets of hypotheses or theoretical frameworks, have been developed to explain and predict the variations in plant defenses, some of which are reviewed by Herms and Mattson [151] and Stamp [341]. The optimal defense (OD) hypotheses state that the plants allocate a proportion of captured resources to defense that maximizes plant fitness. It is assumed that herbivory is a primary selective force and that defenses reduce herbivory [311]. The OD predicts that allocation of energy to defense diverts resources from other needs, such as growth, so that, when herbivory is at a low level, plants not allocating resources to defense will have greater fitness than those that do, as the former will be better competitors in other ways, such as faster growth.

Other theoretical frameworks relate allocation to defense to environmental conditions, such as the influence of carbon and nutrients. One of these, in Bryant et al. [38] challenges the idea that defense should be a positive function of total energy [360]. The authors originally formulated the carbon/nutrient balance (CNB) hypothesis to explain the effects of fertilization and shade on phenotypic variation in secondary metabolism (chemical defenses). The CNB hypothesis predicts that concentrations of carbon-based secondary metabolites will be positively correlated with the carbon/nutrient (C/N) ratio of the plant. What that means is that if the rate of carbon fixation is high compared to the uptake of nutrients, then surplus carbon can be allocated to plant defense. This is likely to happen under conditions of high sunlight and low external input of nutrient. Under shady conditions and/or high nutrient input conditions, the nutrient to carbon ratio in the plant is likely to be high, and it is less likely carbon will be put into carbon-based plant defenses. As nutrient input is decreased and becomes more limiting, it becomes optimal to put a larger fraction of carbon into defense. Another hypothesis, the growth-differentiation balance (GDB) is more comprehensive than the CNB hypothesis and states that any factor that slows growth of the plant to a value lower than what photosynthesis could actually produce will increase the resource pool available for secondary metabolism [150, 151].

Relatively little modeling has been done on this topic. However, the model PLATHO incorporates tradeoff allocations between growth and defense [119]. Those authors used a highly detailed model of herbaceous and woody plant growth, which can be applied to particular plant species under specific abiotic conditions. Gayler et al. [119] incorporated some of the ideas of the CNB and GDB hypotheses into a model to test these against empirical data. PLATHO can be applied in detail to particular plant species under specific abiotic conditions. The authors used the model to determine optimal partitioning of resources in juvenile apple trees, and showed that their results agreed with data on allocation to chemical defenses.

5.2.2.1 PLATHO model

PLATHO (PLAnts as Tree and Herb Objects) simulates the growth of plants, including their carbon and nitrogen fluxes. Its basic processes are morphological development, phenological development, respiration, biomass growth and allocation to biochemical pools, photosynthesis, water uptake, nitrogen uptake, and senescence [120]. An array of submodels is used for the different processes. PLATHO models plant growth and competition in a spatial setting that can hold up to 20 individuals.

The model has been modified from its original form to deal with how carbon and nitrogen are allocated to secondary compounds. The description of the model as applied to allocation of secondary compounds in Gayler et al. [119] is greatly simplified to focus on the behavior of only a few components of the model, including the four variables: A_{av} = available carbon assimilates, W = structural biomass (wood for trees), S = carbon-based secondary compounds, and R = reserves. All are measured in terms of grams of glucose. The model is of hierarchical model type described above, in which fluxes of assimilate to demands by different processes within the plant are prioritized, leading to the set of rules: Resource are allocated for maintenance first, growth takes priority over defense, and assimilates from current photosynthesis will be first used for energy consuming processes before remobilization of reserves is called for. The formation of carbon based-compounds requires sufficient nitrogen levels but is less strongly dependent on nitrogen concentration than on carbon. There is a tradeoff in potential allocation to defensive compounds and plant growth rate.

An equation for the current assimilates can be written, using the authors' notation;

$$A_{av} = (P_{act} + R\tau_R)\Delta t - D_M + A_{old}. \tag{5.14}$$

Here P_{act} is the current photosynthesis rate, τ_R the rate of remobilization of reserves, D_M is the demand for maintenance, A_{old} is the assimilate remaining from the previous time step, and Δt is the time step.

The potential growth rate of total biomass of the plant is

$$D_{pot} = r_{max}W_1\Delta t f_T f_{Ph} \tag{5.15}$$

where r_{max} is the maximum possible growth rate per gram of structural biomass, W_1 is the living structural biomass, and f_T and f_{Ph} are factors between 0 and 1 that depend on temperature and the phenological stage of the plant.

The amount of nitrogen potentially available for growth, N_{av}, is the sum of the uptake from soil and nitrogen remobilization, respectively;

$$N_{av} = N_{upt,pot} + N_{trans,pot}. \tag{5.16}$$

The remobilization during a time step is the summation over actual nitrogen contents over the minimum needed.

$$N_{trans,pot} = \sum_i (N_{act,i} - N_{min,i})\tau_N \Delta t \tag{5.17}$$

where the summation over all of the plant organs, and where $N_{act,i}$ is the current nitrogen content and $N_{min,i}$ the minimal nitrogen content, and τ_N is the rate of mobilization of nitrogen reserves.

It is assumed there is some tradeoff in resources between growth and defense, where $\sigma(0 \leq \sigma \leq 1)$ is the fraction going to defense, and $1 - \sigma$ is that going to structural biomass;

$$D_W = D_{pot}(1 - \sigma) \tag{5.18a}$$

$$D_S = D_{pot}\sigma, \tag{5.18b}$$

where D_{pot}, as defined above, is the maximum potential growth rate of total biomass, and D_W and D_S are the demands for structural growth rate and rate of allocation to defense, respectively. Also, σ is composed of a constitutive part of secondary compounds, σ_C, that is always present in the plant, and an induced part, σ_I, where the latter is greater than zero only when there is actual stress, such as from an attack by a pathogen or herbivore. The latter describes the phenotypic plasticity of the plant with regard to allocation to carbon-based secondary compounds.

The ratios of available carbon assimilate to the potential growth and of available nitrogen to the nitrogen demand needed for that growth are A_{av}/D_{pot} and N_{av}/N_{dem}, where the nitrogen demand, N_{dem}, is determined by the carbon allocation. If these ratios are at least 1, they are set to 1, if they are less than one, they are set to A_{av}/D_{pot} and N_{av}/N_{pot};

$$\phi_C = \min\{1, A_{av}/D_{pot}\} \tag{5.19a}$$

$$\phi_N = \min\{1, N_{av}/N_{dem}\}. \tag{5.19b}$$

The realized growth in structural biomass depends also on the amount of available nitrogen, and is equal to the growth demand times the ratio of available carbon assimilates to potential growth, ϕ_C, times a power, $\beta < 1$, of the ratio of available nitrogen,

$$G_W = D_W \phi_C \phi_N^\beta = D_{pot}(1 - \sigma)\phi_C \phi_N^\beta. \tag{5.20}$$

Given that, from (5.19a), $A_{av} = D_{pot}\phi_C$, then

$$\frac{G_W}{A_{av}} = (1 - \sigma)\phi_N^\beta. \tag{5.21}$$

If the available carbon is greater than what is needed for growth, that is, $G_W/A_{av} < 1$, then the remaining assimilate can be converted to defense compounds, which is the next priority. If the amount of available assimilates exceeds both growth, G_W, and defense demand, D_S, then

$$G_S = \begin{cases} D_S \phi_N^\delta & \text{if } A_{av} \geq G_W + D_S, \\ (A_{av} - G_W)\phi_N^\delta & \text{otherwise,} \end{cases} \tag{5.22}$$

where G_S is the realized amount of defense compound; that is, D_S has been multiplied by ϕ_N^δ to represent the fact that nitrogen is needed in the construction of the defense compounds, so the D_S is multiplied by ratio of available to potentially needed nitrogen, raised to the power δ ($0 \leq \delta \leq 1$). Then also

$$\frac{G_S}{A_{av}} = \begin{cases} \sigma(I)\phi_N^\delta/\phi_C & \text{if } A_{av} \geq G_W + D_S, \\ [1 - (1 - \sigma(I))\phi_N^\delta]\phi_N^\delta & \text{otherwise.} \end{cases} \tag{5.23}$$

If there is a surplus over the amount needed for growth and defense, the rest is given to the reserves.

Gayler et al. [119] used PLATHO with data from apple trees. The results of this analysis and the numerical evaluations of the model that Gayler et al. [119] performed show the following things. First, from equation (5.21) the ratio G_W/A_{av} increases with increasing nitrogen availability and is independent of ϕ_C. But the numerical simulations show that variations of ϕ_C and ϕ_N can have complex effects with respect of allocation to defense-related compounds. The authors' results show that if the plant is low in carbon assimilate, an increase in available nitrogen will decrease contribution to defense, which is in accord with the CNB hypothesis. Near the other extreme, where carbon assimilate is high, an increase of nitrogen availability results in enhanced carbon allocation to defense, as nitrogen is needed in the conversion of assimilates to defense compounds. If nitrogen availability is not sufficient to fulfill the demand for growth ($\phi_N < 1$), this limits carbon allocation to structural biomass. Excess carbon then remains in the pool of available assimilates. If carbon availability is high and nitrogen supply is intermediate, the allocation rate to defense-related compounds can exceed σ. This agrees with the hypothesis that carbon is allocated to secondary metabolism if it cannot be used for growth-related metabolism, which is the GDB hypothesis [151]).

Gayler et al. [119] also simulated the effects of different levels of nitrogen fertilizer on apple trees, which they compared with data. They found that increased growth rates resulting from high fertilization were associated with decreased levels of defense compounds in leaves, which agrees with the CNB hypothesis.

5.2.3 Methods of compensation for herbivore damage

Plants are able to respond to herbivory by increased levels of defense. But they also have compensatory mechanisms that can reduce the effects of herbivory. Some of these arise simply from the structure of the plant. Leaves, if produced in large number, will shade each other, eventually to the point that the bottom leaves don't get much sunlight. Consider some agent of defoliation, such as an herbivore, which reduces a tree's foliage by some amount. Although this reduces the surface area of foliage available to capture light, removal of foliage does not necessarily reduce light capture, and hence potential growth, in a linear way. This can be seen from one of the simplest models representing the relationship between leaf foliage and the growth rate, G;

$$G = G_0\left(1 - e^{-kF}\right) \tag{5.24}$$

where G_0 is the maximum possible rate of growth based on photosynthesis, F is the amount of foliage, that is the leaf area index, and k is the rate of extinction of light passing through the foliage. The terms $1 - exp(-kF)$ represent the fraction of available light captured as a saturating response. If F is initially very large, such that $1 - exp(-kF)$ is very close to 1, even reducing F by one half might not appreciably reduce the light captured. This is a compensatory mechanism reflecting the fact that trees commonly have much foliage that is relatively shaded, which can compensate for the loss of foliage above it by being exposed to a higher level of radiation.

A second compensatory mechanism is the way the tree can change the way it allocates its resources in response to losses from disturbances such as herbivory. Reduction of shoot material, for example, may be compensated for by the plant producing proportionally more shoot and less root material [57, 58]. This allocation can be adjusted to be more appropriate to the assumed added level of foliage removal. In particular, the allocation of these resources can be altered to increase those going to foliage by reducing the amount going to roots and/or to wood. The reason reallocation from roots to foliage might be effective in some cases is that the tree's current capacity for water and nutrient uptake through roots may be in excess of its needs. Just as a plant can have more foliage than it needs to capture a certain amount of radiation, it may have more roots than it needs at the time, so transferring a greater fraction of resources from roots to foliage may come at not too high a cost to nutrient uptake. An additional compensatory effect is that, as a tree reduces its uptake of nutrients from the soil, nutrient concentration in the soil can increase, partly offsetting the reduction in intensity of exploitation. Therefore, it is necessary to study the effects of foliage removal by considering the whole plant and soil as a system, in which the plant can alter the way it alters resource allocation in an optimal way in response to control measures such as defoliation.

Of course, some of the tree carbon and nutrient resources could also be reallocated to anti-herbivore defenses; so, if a herbivore consumes foliage, the plant may respond by chemical defenses, as well as by reducing the nutrient:carbon ratio in its leaves to diminish the effectiveness of the agent. We will not consider this factor here, but only attempt to find the best response of the plant in terms of changes in the way it allocates carbon and nutrient, when defoliation is increased. The best strategy for success will be assumed to be not simply a repair of the removed foliage, but allocation of resources among its components (foliage, roots, and stem) in a way that maximizes growth rate. Then the question can be asked, how the tree will respond to an increased rate (assumed constant here) of loss of leaves, in particular, how it might change its allocations to foliage and roots in an optimal way.

5.2.3.1 Model of plant compensation

The dynamics of a tree in steady state is modeled to determine the allocation strategy that maximizes growth under given conditions of light, nutrient availability, and defoliation. To do this, a well-known model of tree growth and nutrient cycling; the G'DAY model [72], is used. This model simulates both carbon and nitrogen in tree and soil compartments. Ju and DeAngelis [179] used a modified version of this model that was enhanced by adding an explicit compartment for soil pore nutrient (Figure 5.11). However, to avoid soil process complexity the seven compartments for litter and soil in the original G'DAY model were removed and mineralization of nutrient from litter was assumed to occur instantaneously, going straight to the soil pore water, without being processed through all of the litter and soil compartments. It was also assumed that some nutrient could be lost during recycling at a rate proportional to its flux through the plant biomass. This model has an advantage that it can be used to study the effects of different allocation strategies by trees of energy to foliage, roots, and wood.

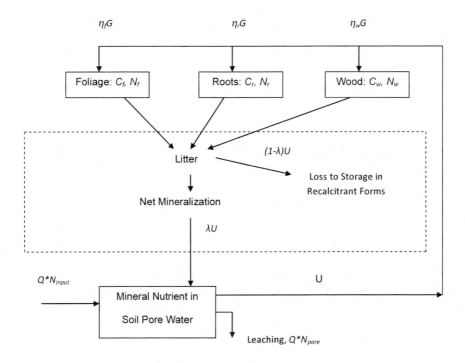

FIGURE 5.11: Schematic of model for allocation of carbon and nutrient among three compartments of woody vegetation; foliage, roots, and wood.

With the above assumptions the G'DAY model reduces to six equations for carbon and nutrient (e.g., nitrogen or phosphorus) in the three tree components, foliage (C_f and N_f), fine roots (C_r and N_r), and wood (C_w and N_w);

$$\frac{dC_f}{dt} = \eta_f G - \gamma_f C_f \tag{5.25a}$$

$$\frac{dC_r}{dt} = \eta_r G - \gamma_r C_r \tag{5.25b}$$

$$\frac{dC_w}{dt} = \eta_w G - \gamma_w C_w \tag{5.25c}$$

$$\frac{dN_r}{dt} = (U - \eta_w \nu_w G)\frac{\eta_f}{\eta_f + \rho\eta_r} - \gamma_f N_f \tag{5.25d}$$

$$\frac{dN_f}{dt} = (U - \eta_w \nu_w G)\frac{\rho\eta_r}{\eta_f + \rho\eta_r} - \gamma_r N_r \tag{5.25e}$$

$$\frac{dN_w}{dt} = \eta_w \nu_w G - \gamma_w \nu_w C_w, \tag{5.25f}$$

plus a seventh equation for the soil pore water nutrient concentration, N_{pore};

$$\frac{N_{pore}}{dt} = Q(N_{input} - N_{pore}) - U + \lambda(\gamma_f N_f + \gamma_r N_r + \gamma_w \nu_w C_w). \tag{5.25g}$$

Separate equations for carbon and a limiting nutrient allow the N:C ratio in foliage, roots, and wood to be variable. The constants γ_f, γ_r, and γ_w are senescence (i.e., litterfall) rates. It is assumed that a fixed ratio, ν_w, of N to C, is first allocated to wood, and then the rest of the nutrient is allocated to foliage and fine roots in the proportions $\eta_f/(\eta_f + \rho\eta_r)$ and $\rho\eta_r/(\eta_f + \rho\eta_r)$, respectively.

In the above equations, the function G represents net carbon production, or growth rate. In particular,

$$\begin{aligned} G \quad &= \text{net carbon production, or growth per unit time (kg m}^{-2}\text{yr}^{-1}) \\ &= G_0 I(C_f) E(\nu_f) \end{aligned} \tag{5.26}$$

$$\begin{aligned} I(C_f) \quad &= \text{light interception factor} \\ &= 1 - exp(-k_f b_f C_f) \end{aligned} \tag{5.27}$$

$E(\nu_f)$ = rate-limiting effect of low nutrient concentration on growth, where

ν_f =N:C ratio in foliage = N_f/C_f

$$E(\nu_f) = \nu_f/(\nu_0 + \nu_f). \tag{5.28}$$

The parameter G_0 is the maximum possible primary production, b_f is the foliage per unit carbon, and k_f is the light extinction factor. The factor $E(\nu_f)$ represents the assumption that the photosynthetic efficiency of foliage is dependent on the N_f:C_f ratio. The three parameters, η_f, η_r, and η_w, govern the allocation of energy between foliage, fine root biomass, and wood, respectively, where $\eta_f + \eta_r + \eta_w = 1$. To keep the present analysis simple, we assume here that η_w is fixed; that is, whatever the relative allocations to foliage and roots, the fraction allocated to wood stays the same.

The function U represents nutrient uptake,

U = uptake rate of plant-available nutrient (kg m^{-2} yr^{-1}), where we assume a saturated response of uptake to pore water concentration, and also that there is a "resource extinction" rate, k_r, that multiplies fine root biomass, analogous to light extinction. The parameter k_r multiplies a coefficient of fine root length per unit carbon, b_r, and the amount of carbon in fine roots, C_r (see [148]):

$$U = \left(\frac{g_N N_{pore}}{k_N + N_{pore}} \right) \left(1 - e^{-b_r k_r C_r} \right), \tag{5.29}$$

where g_N is the maximum possible nutrient uptake rate and k_N is the half saturation constant. In equation (5.25g),

$Q = $ flow of water through the soil (kg m^{-2} yr^{-1})

$N_{input} = $ nutrient concentration in input water (g nutrient g^{-1}water)

$\lambda = $ fraction of nutrient recycled; the remainder is assumed tied up in recalcitrant

forms or, if nitrogen, also lost to gaseous forms.

If some loss of available nutrient to recalcitrant forms or loss to the atmosphere potentially occurs during decomposition of litter, then $0 \le \lambda \le 1$. However, it is conceivable that $1.0 < \lambda$. This could happen if the vegetation captures external nutrient (say by "combing" it from the atmosphere, or extracting it from ground water) at a rate that is linearly proportional to the standing stock of biomass and that exceeds the losses to recalcitrant compounds. We are neglecting the dynamics of water, assuming that it is present in sufficient amount to facilitate nutrient uptake under steady-state conditions. All parameters are defined in Table 5.2.

It can be shown that, for a given maximum possible photosynthetic rate, G_0, and nutrient input to the system, N_{input}, a maximum value of growth rate, $Max(G)$, exists for some combination of the three allocation parameters η_f, η_r, and η_w where $\eta_f + \eta_r + \eta_w = 1$. Because η_w is fixed, only η_f and η_r vary and η_r is a function of η_f. Therefore, refer to the strategy that maximizes G as $Optimal(\eta_f)$. Given $Max(G)$, and the allocation fractions, it is possible to find all of the tree components, foliage, roots, and wood that result from this objective. In addition, the analysis also can determine the strategy that maximizes the amount of foliage, which is given by $Max(\eta_f G)$, and the strategy that minimizes N^*_{pore}.

5.2.3.2 Deriving a function for the optimization of G

Here it is calculated how a tree should allocate its energy (or carbon) resources in a way that maximizes its growth rate, G, in the absence of competition. The above model is analyzed to find the maximum value of G as a function of the three allocation parameters η_f, η_r, and η_w where $\eta_f + \eta_r + \eta_w = 1$. In particular, this can be explored for the effect of different N_{input} concentrations on the maximum value.

Assume that the system is at steady state, so that each of the equations (5.25a-g) is set to zero. It was not possible to find an explicit analytic expression for G, from which a maximum of G could be determined analytically, but it was possible to do so numerically from an implicit expression. Our first objective was to find an expression for G, which resulted in an implicit equation for G,

$$G = R_0 \left(1 - e^{-k_f b_f \eta_f G/(\gamma_f + \phi_f)} \right) \left[\frac{v_f}{v_f + v_0} \right], \tag{5.30a}$$

where

$$v_f = \left[\left(\frac{g_N N_{pore}}{k_N + N_{pore}} \right) \left(1 - e^{-k_r b_r \eta_r G/\gamma_r} \right) - \eta_w v_w G \right] \left(\frac{\eta_f}{\eta_f + \rho \eta_r} \right) \left(\frac{1}{\eta_f G} \right) \tag{5.30b}$$

TABLE 5.2: Variables and parameters used in the model.

Variables	
C_f, C_r, C_w = carbon pool for foliage, root, and wood (kg m^{-2})	
N_f, N_r, N_w = nitrogen pool for foliage, root, and wood (kg m^{-2})	
N_{pore} = soil pore water nitrogen pool (kg nutrient kg^{-1} water)	
η_f, η_r = allocation fraction of carbon to foliage and root	

Parameters	Values(s) or Range(s)
η_w = allocation fraction of carbon to wood	0.40
ρ = ratio of root N:C to foliage N:C ratio (assumed constant)	0.70
ν_w = N:C ratio for wood	0.003
γ_f = senescence rate for foliage (yr^{-1})	variable
γ_r = senescence rate for root (yr^{-1})	0.60
γ_w = senescence rate for wood (yr^{-1})	0.02
λ = recycling ratio	0.98
G_0 = maximum growth rate (dry mass basis) (kg m^{-2}yr^{-1})	7.0
k_f = light extinction coefficient	0.5
b_f = specific leaf area (dry mass basis) (m^{2}kg^{-1})	10.0
b_r = root length per unit C	0.5
k_r = soil resource extinction	0.1
g_N = maximum possible steady nutrient uptake rate per ground area	3.00
k_N = half-saturation coefficient for N plant uptake	0.00004
N_{input} = input concentration of nutrient (kg (kg water)$^{-1}$)	0.000001
Q = flow of water (kg m^{-2} yr^{-1})	100

and

$$\eta_r = 1 - \eta_f - \eta_w.$$

This equation still contains N_{pore}, which is a variable, so N_{pore} must next be eliminated in terms of G in order to obtain an equation that contains only the variable G. We did that by using the right-hand sides of equations (5.25d,e,g), along with equation (5.25c). These allowed us to eliminate N_f, N_r, and C_w, so that N_{pore} is expressed simply in terms of G in the second-order equation

$$QN_{pore}^2 + \left[g_N(1-\lambda)\left(1 - e^{-b_r k_r \eta_r G/\gamma_r}\right) + Q(k_N - N_{input})\right]N_{pore} - QN_{input}k_N = 0. \quad (5.31)$$

Solving this for N_{pore} yields,

$$N_{pore}^* = \frac{-B \pm (B^2 + 4Q^2 N_{input} k_N)^{1/2}}{2Q} \quad (5.32a)$$

where

$$B = g_N(1-\lambda)\left(1 - e^{-b_r k_r \eta_r G/\gamma_r}\right) + Q(k_N - N_{input}). \quad (5.32b)$$

It can be shown that the only relevant solution of (5.31) is the one with the plus sign in (5.32a). Then N_{pore} can be plugged into (5.30b) and (5.30b) into (5.30a) to obtain the final implicit equation for G.

5.2.3.3 Numerical evaluation of G as a function of allocations η_f, η_r, and η_w

Equations (5.30) and (5.32a,b) were evaluated numerically for G for the baseline set of parameters shown in Table 5.2 and different values of N_{input}, and with $\lambda < 1$. To simplify the analysis, we set η_w to some particular value, by assuming that the energy allocation to wood is constrained at some level needed to support both foliage and fine roots.

To determine how the optimal allocation fraction to foliage, η_f, changes as a function of defoliation, the loss rate of foliage, γ_f, was varied from 0.3 to 1.05. Over this interval, η_f increased from 0.34 to 0.45 (Figure 5.12a). The values of G attained for all possible values of foliage allocation, η_f, are shown for three values of the defolation rate, $\gamma_f = 0.3, 0.65$, and 1.05, are shown in Figure 5.12b. Note that the maximum value of G, which is $Max(G)$ decreases with increasing defoliation. This translates into decreases in carbon in foliage, fine roots, and wood, as the rate of foliage loss is increased. $Max(G)$ decreases as the defoliation rate, γ_f, is increased (Figure 5.12c). However, note that, although γ_f is tripled in value, $Max(G)$ declines only by one-half, because of compensatory effects in the plant.

The implication of these results is that plants can use their compensatory capacity to lessen the effects of herbivory. In the model presented here, herbivory always does have a negative effect, though less drastic than the level of herbivory would suggest. However, there is empirical evidence of herbivory not having a negative effect, but possibly having a positive effect on primary productivity of plants [154, 253]. The likely mechanisms for such overcompensation are not known, but some possible explanation has been given by de Mazancourt et al. [80]. In theoretical models, they showed that if several conditions are satisfied, herbivory could increase primary production up to a moderate rate of grazing intensity through recycling of a limiting nutrient. These models will not be described here.

5.3   Glossary

Carbon/nutrient balance model: This hypothesizes that when carbon assimilates are in stoichiometric excess in a plant, the plant will tend to allocate these to carbon-based chemical defenses, such as phenolics, whereas, when nitrogen is in stoichiometric excess, the plant will tend to allocate this to nitrogen-based defenses, such as alkaloids.

Constitutive chemical defense: This defense is always present in a particular plant species.

Convergence stable strategy: An evolutionarily stable strategy (ESS) is also convergence stable, if any strategy of an invader that is closer to the ESS than the current strategy of the resident will win over the resident's strategy.

Evolutionarily stable strategy (ESS): An ESS is a strategy that, if adopted by a population, cannot be invaded by any other strategy that is initially rare.

Functional equilibrium model: This hypothesizes that the allocation of carbon and nutrients serves the purpose of optimizing some goal such as maximizing carbon production under given environmental conditions.

Growth-differentiation balance model: This model distinguishes growth (production of leaves, roots, and stems) from differentiation; the latter can include production of secondary metabolites and other defense measures. It hypothesizes that if some factor limits growth relative to photosynthesis, this will increase the resources that can be allocated to defenses.

Ideal free distribution (IFD): An IFD is a strategy originally applied to the distribution of animals in spaces. It assumes that the animals are free to move and that they will distribute themselves in space such that no individual can improve its fitness by

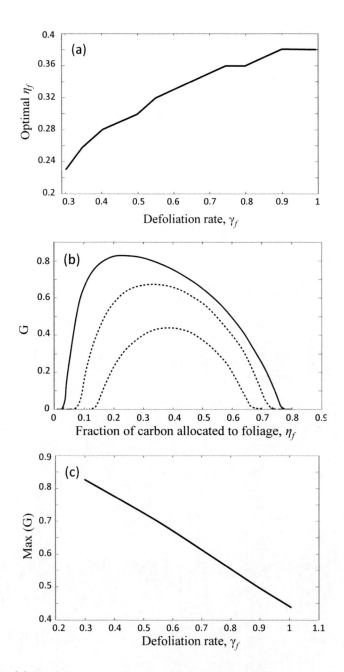

FIGURE 5.12: (a) Model output of optimal allocation to foliage, η_f, as a function of the defoliation rate, γ_f. It is assumed that $\gamma_f = 0.3$ is the baseline loss of foliage, with higher levels resulting from herbivory. (b) Profile of growth rate as a function of allocation to foliage, η_f, for three values of the defoliation rate γ_f, $\gamma_f = 0.3$, 0.65, and 1.05. The peaks of the curves occur for $Optimal(\eta_f)$ and given $Max(G)$. (c) The maximum attainable growth rate, $Max(G)$, for optimal allocation of carbon, as the defoliation rate, γ_f, increases. Note that although the defoliation rate almost triples, the $Max(G)$ is only reduced by about one-half.

moving. This concept has been applied to other characteristics of a species population as well.

Inducible chemical defense: This type of chemical defense is only present in the plant when an attack by herbivores has occurred or, in some cases, when the presence of an herbivore attack nearby is detected.

Optimal defense model: This model hypothesizes that plants will allocate resources between defenses and other needs, such as growth, in a way that optimizes fitness.

Transport-resistance model: This assumes that carbon assimilates in a plant flow along concentration gradients from the source areas where they are created (e.g., leaves) to places where they are used for growth and other processes.

Chapter 6

Herbivore Strategies: The Role of Plant Quality and Defenses

6.1   Herbivore strategies in exploiting plants

6.1.1   Introduction

The effects of herbivory on terrestrial primary productivity, especially that of forests, are generally small [291]. Herbivores are seldom able to exploit more than a fraction of the primary productivity in terrestrial ecosystems [291]. The fact that a lot of plant biomass is uneaten by herbivores is sometimes called the "world is green" phenomenon [297, 383]. Herbivores are generally kept in check by some combination of factors, including:

1. Properties of the individual plant such as low nutrient concentrations and defenses that include morphology and chemicals

2. The complex of predators and pathogens attacking the herbivore

3. Abiotic factors such as weather

While plant strategies are the focus of most of this book, herbivores certainly respond to maximize their fitness under the given conditions. Foraging theory analyzes herbivore foraging strategies by which the herbivores address the above challenges, including their response to low plant quality and chemical defenses. The objective here is not to review this field, which has a huge literature (e.g., [63, 106, 116, 236, 250, 280, 297, 300, 327, 332, 338, 343, 381, 382]). Instead we point to a few representative modeling approaches.

We first briefly review properties of the plant, both the low ratios of nutrients in plant biomass and the use of chemical defenses, and then describe models of strategies of herbivores to maximize their fitness in spite of the challenges faced in exploiting plants. Two aspects of chemistry are of interest here, food quality and defense chemicals.

6.1.1.1   Food quality

As noted in Section 3.6, plant tissue is composed mostly of cellulose and lignin, while nitrogen concentrations in mature foliage rarely exceed 2%. Because the bodies of herbivores such as insect larvae contain more than 10% nitrogen, this difference affects the efficiency of conversion of ingested plant food; that is the assimilation efficiency. In particular, efficiency of conversion of ingested food varies with nutrient content. The effect of the ratio of "energy to protein" on pupal weight of insects feeding on a variety of diets is clear from experimental data; a high "energy to protein" ratio, with its empty calories, leads to lower body size of pupae [328].

6.1.1.2   Plant defenses

Plants possess morphological defenses like thorns and trichomes that limit the ability of some herbivores to feed on plant biomass. By the 1950s, knowledge about the amount, array, and biological activity of "secondary metabolites" in plants suggested that these compounds also served as defense against pathogens and herbivores [341]. Tens of thousands of such chemicals have been described. Herms and Mattson [151] consider the question of allocation between growth and defense to be a central dilemma of plants. The effect of chemical defenses is a form of feedback regulation, because the defenses can slow down the herbivore growth and reproduction or even kill them, thus regulating their population sizes.

6.1.2   Herbivore adaptations

What can herbivores do to compensate for poor plant quality, that is, low nutrient to biomass ratio, and plant chemical defenses? This topic has attracted much attention in the literature, including important reviews (e.g., [109, 112, 183, 312]). Herbivores are adapted in many ways to extract the nutrients they need. Karban and Agrawal [183] divided the herbivore behavior strategies of dealing with properties of plants into three levels, least aggressive, sometimes aggressive, and aggressive. The least aggressive strategies include feeding choices herbivores make, e.g., when, where, and on what plant species to feed, and oviposition choices by female herbivorous insects. The authors classify strategies as sometimes aggressive, for example, when they involve producing enzymes that are able to reduce the detrimental effects of plant defense compounds, sequester these compounds, or use microbial symbionts to improve their ability to exploit their host plants. Aggressive strategies are classified as those in which plants are manipulated in ways that allow them to be more effectively exploited. For example, some aphids create galls in plants, which they can inhabit and feed on the plant. Other insects clip leaf veins to prevent the plant from responding with anti-herbivore exudates, or use gregarious feeding to overcome plant defenses.

White [383] noted a number of ways that herbivores, faced with an autotrophic level with generally low nutrient to biomass concentrations, are able to either find transitory concentration hotspots in space and time, or have mechanisms to extract needed nutrients from the available low concentrations. Some of the specific techniques are the following.

Flush-feeding. This is eating only the new growing leaves of food plants, as young leaves contain higher concentrations of nutrients. Koalas, for example, eat nothing but new growth of gum trees. Aphids, for example, tend to feed near the tips of growing stems and flower buds.

Seed-eating. Many herbivores, such as white cockatoos, eat only the unripe seeds ("milk-ripe seeds") of plants, which contain amino acids in soluble form. Ripe seeds have relatively indigestible protein.

Grazing the same plants over and over again. This keeps some plants in a continually growing state that can be used by flush feeders. Giraffes do this with acacia trees so that they can always be feeding on young growth.

Creaming off the best. The herbivore passes a lot of food through its gut, but skims off only the easily available nitrogen, leaving the harder to obtain nitrogen. Pandas do this with bamboo, which they crush finely, and pass through the gut in 8 hours, extracting only the easily available nitrogen.

Passing huge quantities through the gut. Spittle bugs feed on very dilute xylem sap. They ingest 150 to 250 times their body weight in a day. They tend to feed on legumes, which have relatively high (compared with other plants) organic nitrogen in their xylem (from their N-fixing bacteria).

Senescence-feeding. These are sap-sucking insects that feed on dying leaves from which high concentrations of nutrient are being translocated by the plant.

Coprophagy. This can be feeding by herbivores on the feces of carnivores, which is relatively high in protein. Some herbivorous fish do this. Hind-gut fermenters such as voles, koalas, hares, rabbits, grouse, carry bacteria in specialized caeca, where pellets of concentrated microbial protein are formed. In some cases, for example hare, these are eaten.

Fore-gut fermenters. Grazers like cows and kangaroos have specialized stomachs where microbes can digest the cellulose that the herbivores cannot digest themselves.

Digesting microbes. Termites and cockroaches that live on wood harbor microbes that can digest the cellulose, and, in the case of the termites, fix nitrogen.

Detritus feeding. This is feeding on the microbes that are decomposing the detritus. Wood lice do this.

Meat-eating "vegetarians." Many herbivores thought to be pure vegetarians will eat meat when they can. These include rock wallabies, dugongs, squirrels, gorillas, pandas.

The above are examples of specific strategies of certain types of herbivores. A widely practiced strategy of generalist herbivores is to feed on a mixture of plants to obtain sufficient energy and the right amounts of nutrients, and to avoid high accumulations of toxins from any particular plant species. We examine a model that illustrates this.

6.1.3   Modeling feeding choice to optimize intake of a limiting nutrient

Phytophagous insects tend to be specialists, or monophagous (feed on one plant species, genus, or family). Dan Janzen speculated that at least one-half of the folivorous insect species in the tropics consume only one species of plant, and none more than 10% of available plants [172]. In temperate and boreal systems there are numerous examples of insects that have distinct preferences. Some herbivorous insects will feed on more than one species; e.g., western spruce budworm prefers true firs and Douglas fir, but will eat pines and larches. But many are more specific–sawflies, bark beetles, tend to specialize.

Vertebrate herbivores tend to be generalists (polyphagous), however, so, although they may have preferences, they can feed on many plant types. (We will use the term "plant types" here. It may mean different species, but could refer to many other differences in the plant biomass consumed, including biomass from different parts of the same plant.) This gives herbivores the option of being selective when they encounter a variety of plant types. Models of vertebrate, particularly ungulate, feeding strategies have been constructed. We will describe a few. We start with the model of Owen-Smith and Novellie [279], called a "contingency model," which was applied to an African ungulate, the kudu (*Tragelaphus strepsiceros*).

Recall that in Chapter 2 we derived an expression for the rate of intake P_c (e.g., grams per day) of prey biomass (plant in this case) when the predator (herbivore in this case) has the choice of two plant types;

$$P_c = \frac{\sigma_1 f_1 P_1 + \sigma_2 f_2 P_2}{1 + \sigma_1 f_1 h_1 P_1 + \sigma_2 f_2 h_2 P_2}. \tag{6.1}$$

Here $f_i = 2vk_is$ is the area that the herbivore searches during a time period, where s is the mean velocity with which the herbivore moves, and v is the herbivore's plant sensing range on either side of its patch, and k_i is prey detectability, which will differ for different prey types. Also, we introduce the assumption that the predator can make a choice of whether to attack a particular prey encountered; it may prefer one prey type to the other. So we introduce the parameters, σ_1 and σ_2 to represent the probabilities that the predator will attack a prey of a given type; that is, these are selectivities.

Expression (6.1) can easily be generalized to the case where there are n plant types that the herbivore may choose among;

$$P_c = \frac{\sum_{i=1}^{n} \sigma_i f_i P_i}{1 + \sum_{i=1}^{n} \sigma_i f_i h_i P_i}. \tag{6.2}$$

These n plant "types" are not necessarily different species. They may be different plant parts of the same species, or they may be different functional plant types. To put this expression in a form that can make use of the parameters supplied by Owen-Smith and Novellie [279] for the kudu, we make certain changes. We represent the rate at which vegetation is encountered by $h \times w \times k$, where

> h = the effective height reach of the animal (m)

> k = the effective patch width scanned (m)

> w = the walking rate (m min^{-1})

In addition, we replace the handling time, h_i, by its equivalent, $1/e_i$, where e_i is the eating rate of the plant biomass by the herbivore (g min^{-1}). Also, to take into account that the herbivore does not spend all of its time searching for and handling plant resources, we multiply P_c by the time actually occupied by searching and feeding $(1 - S)$. Finally, the objective of the model of Owen-Smith and Novellie [279] is to determine the maximum possible ingestion of a limiting biomass component, protein, rather than biomass, P, so P_i, defined here as biomass per square meter, is replaced by $c_i P_i$ (g protein m^{-2}) in the numerator of expression (6.2), where c_i is the ratio of the protein to biomass. The denominator remains as is, because ingestion is still assumed to saturate with respect to biomass, not protein.

So the equation for nutrient (protein) ingestion rate, NIR (g min^{-1}) now becomes

$$NIR = \frac{hkw \sum_{i=1}^{n} \sigma_i P_i c_i}{1 + hwk \sum_{i=1}^{n} \sigma_i P_i / e_i}(1 - S) \tag{6.3}$$

where the multiplier $(1 - S)$ is included to take into account the authors' estimate of the time the kudu devotes to feeding activities. This expression differs from the expression for NIR in [279] only in that we use σ_i where they used a_i, and we use P_i where they used d_i.

Owen-Smith and Novellie [279] assumed that the biomass densities of the plant types differed and that also the protein to total biomass ratio varied as well. Plant types with high ratios of protein were the rarest, with plant types with lower ratios being increasingly common.

The question that Owen-Smith and Novellie [279] ask is: What is the choice of values

TABLE 6.1:   Parameter values for the model of Owen-Smith and Novellie [279].

$h = 2.5$ (m)
$k = 0.5$ (m)
$w = 0.33$ (m sec^{-1})
$d_i = 2, 3, 5, 7.5, 10, 15, 22, 33, 50, 75$ (g m^{-1})
$e = 10$ (g min^{-1})
$S = 0.15$
$c_i = 0.24, 0.22, 0.20, 0.18, 0.16, 0.14, 0.12, 0.10, 0.08, 0.06$ (g protein g biomass^{-1})

of σ_i that maximizes NIR? This is easily found by evaluating all possible combinations of σ_i, each of which can vary from 0 to 1. An important assumption of the model is that σ_i can take only the binary values 0 and 1. Therefore, a plant type is either consumed or not. More important is that there is no restriction on the σ_i resulting from the value that any other σ_i takes; that is, the herbivore is assumed to search for all plant types simultaneously.

Given the assumption of the model and the parameter values in Table 6.1, the results are shown in Figure 6.1. The x-axis is the cumulative number of plant types exploited, starting from the rarest (and most protein rich) to the most common (and protein poorest). Figure 6.1 shows the protein ingestion peaks at an intermediate number of accepted species.

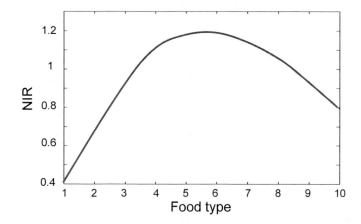

FIGURE 6.1: Prediction of protein intake (NIR) as a function of the cumulative number of plant types that are selected by the foraging herbivore.

To optimize its nutrient intake, the herbivore should reasonably consume the plant types encountered that have the higher ratios of protein to biomass, the first six types in this case, but then reject the lowest quality plant types, even though they are the most frequently encountered. The latter deliver little protein but, if consumed, would monopolize the herbivore's gut to the exclusion of the high-quality plant biomass. This selection of only the highest-quality plant types is a consequence of the assumption that the herbivore is able to search for all plant types simultaneously. The fact that the herbivore encounters low-quality plants more frequently than high-quality plants does not diminish the rate at which the herbivore encounters the latter. Also, the herbivore is assumed to have perfect knowledge of plant quality. These assumptions are usually not true in nature, of course, which is why in empirical studies herbivores are found to have a wider diet than contingent models show. Also, as Pastor [284] has pointed out, herbivores such as moose actually change the composition of the plant community through time, so the model of Owen-Smith and Novellie [279] is strictly valid only for describing short-term optimal strategies. Over longer terms, selectivity values may have to change to reflect changes in plant community composition.

6.1.4   Constraint on plant toxin intake

Real situations are complex, and consideration of only one nutrient may not be sufficient to predict proportions of plant types consumed. Owen-Smith and Novellie [279] argue reasonably that high levels of protein will often be associated with high levels of other key nutrients. But other factors may also affect the exploitation of particular plant types, one important factor being the toxin content of biomass. One assumption that can be made is that there is a threshold value of toxin ingestion that the herbivore can tolerate, but in no case wants to exceed. This adds a constraint to the model

$$\sum_{i=1}^{n} Tox_i \leq Tox_{limit} \tag{6.4}$$

where tox_i is the toxin ingested from each plant type and Tox_{limit} is the limit of intake over a particular time period beyond which the herbivore will not go. In modeling the effect of this constraint here, it will be assumed for simplicity that the type of toxin is the same in each plant type, so that the toxin from the different plant types is additive.

We do not have data on the toxin contents of the plant types in the model. However, for illustrative purposes, assumptions can be made to show how this constraint affects the selection of plants. Suppose that the plants contain toxins in proportion to their nutritional value. Let the ratio of plant toxin to plant protein be tox_i, so the amount of toxin per unit biomass is assumed to be $tox_i c_i P_i$. Then the rate of intake of toxin, TIR, is

$$TIR = \frac{hkw \sum_{i=1}^{n} \sigma_i c_i P_i tox_i}{1 + hwk \sum_{i=1}^{n} \sigma_i P_i / e_i}(1 - s). \tag{6.5}$$

If we require that $TIR < Tox_{limit}$, this restricts the set of values of σ_i that can be used by the herbivore.

As an example, unlike the case described above, we now allowed σ_i for each species to have a range of possible values from 0 to 1; the possible values being 0, 0.2, 0.4, 0.6, 0.8, and 1.0. We maximized (6.3a) subject to the condition $TIR < Tox_{limit}$ to examine how different values of Tox_{limit} affect the maximum possible NIR (from (6.3)). We allowed tox_i to be the same, $tox_i = 0.2$ for all values of i, and considered a sequence of values of Tox_{limit} from 0.15 to 0.35.

The results show that, as Tox_{limit} increases, NIR increases and reaches a maximum at $NIR = 1.171$ for $Tox_{limit} = 0.25$ (Figure 6.2), which it remains at for all higher values of Tox_{limit}. This is the same as the maximum NIR in Figure 6.1. For values of $Tox_{limit} \geq 0.25$, the array of values of σ_i has the form shown in Figure 6.3a. For values of $Tox_{limit} < 0.25$, the array of values of σ_i will be different, shifting from more toxic to less toxic plant types; e.g., Figure 6.3b shows the array of values of σ_i for $Tox_{limit} = 0.17$. This example shows how constraints can have major effects on the herbivore's selections of plant types to consume.

6.1.5   An herbivore foraging on spatially distinct patches

The above model applies to an herbivore feeding in a single patch. This meant that the herbivore could search for all plant types simultaneously, and so, with perfect information on relative rewards of plants, had the ability to maximize intake by feeding on the highest

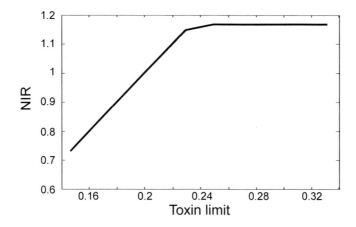

FIGURE 6.2: Prediction of the maximum intake of protein that can occur as the upper limit on the level of toxins that herbivore can ingest is increased. When this limit exceeds 0.25, the toxin has no effect on the ability of the herbivore to maximize its intake of protein.

quality plants. This advantage is lost when plants of different types are separated, so that searching for one type of plant reduces the time that can be spent searching for other types.

Consider, alternatively, that more than one patch exists and that the herbivore can move between them. For simplicity, assume there are only two patches and that each contains only one species of plant. It can be assumed that it takes time to move between patches. If a mean amount of time T_{trav} per unit time is taken moving between the patches, then, from the considerations of Chapter 2, it can be shown that the mean rate of biomass intake, E, has the form

$$E = \frac{\sigma_1 f_1 P_1 T_s + (1 - \sigma_1) f_2 P_2 T_s}{T_s + \sigma_1 f_1 h_1 P_1 T_s + (1 - \sigma_1) f_2 h_2 P_2 T_s + T_{trav}} \tag{6.6}$$

where it is also recalled that T_s is the mean fraction of time spent searching. This expression is due to Fryxell and Lundberg [116], who then assumed travel time to be small, so that the expression is simplified to

$$E = \frac{\sigma_1 f_1 P_1 + (1 - \sigma_1) f_2 P_2}{1 + \sigma_1 f_1 h_1 P_1 + (1 - \sigma_1) f_2 h_2 P_2}. \tag{6.7}$$

The selectivity σ_2 has been replaced by $1 - \sigma_1$ here, because feeding in one patch necessarily prevents feeding in the other at the same time. It is possible to find the optimal value of σ_1 by taking the derivative of E with respect to σ_1;

$$\frac{\partial E}{\partial \sigma_1} = \frac{f_1 P_1 (1 + f_2 h_2 P_2) - f_2 P_2 (1 + f_1 h_1 P_1)}{(1 + \sigma_1 f_1 h_1 P_1 + (1 - \sigma_1) f_2 h_2 P_2)^2}. \tag{6.8}$$

For E to increase with increasing σ_1, it is necessary that $\partial E / \partial \sigma_i > 0$ or

$$\frac{f_1 P_1}{1 + f_1 h_1 P_1} > \frac{f_2 P_2}{1 + f_2 h_2 P_2}, \tag{6.9}$$

which means that the benefits of foraging in patch 1 outweigh those of foraging in patch 2. Because σ_1 does not occur in this inequality, it means that this holds for all values of σ_1, so that E is maximized when σ_1 takes its maximum possible value of 1.

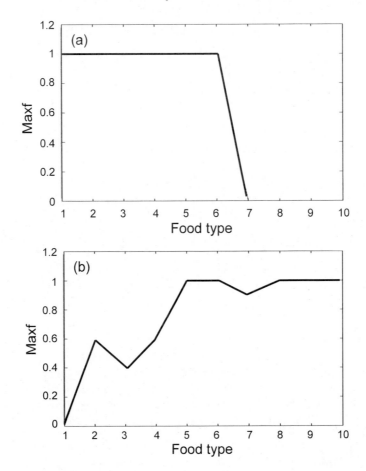

FIGURE 6.3: The predicted fractions of encountered plant types the herbivore is able to ingest when the upper limit on toxin that (a) can be ingested exceeds 0.25, or (b) is less than 0.17, and thus less than the threshold of 0.25.

The situation modeled here becomes more interesting if there is also some risk, for example, from natural enemies. This can be done by assuming that there is some level of risk per unit time for the foraging being in either patch, which differs for the two patches. Fortin et al. [106] assume that staying in a particular region also presents risk of predation, quantified as u, which can be represented as

$$u = \frac{\sigma_1 u_1(1 + f_1 h_1 P_1) + (1 - \sigma_1)u_2(1 + f_2 h_2 P_2)}{1 + \sigma_1 f_1 h_1 P_1 + (1 - \sigma_1)f_2 h_2 P_2}. \tag{6.10}$$

Note that the term $\sigma_1 u_1(1 + f_1 h_1 P_1)$, for example, represents the risk over a unit of time associated with the sum of the searching time and prey handling time spent in patch 1. Then

$$\frac{E}{u} = \frac{\sigma_1 f_1 P_1 + (1 - \sigma_1)f_2 P_2}{\sigma_1 u_1(1 + f_1 h_1 P_1) + (1 - \sigma_1)u_2(1 + f_2 h_2 P_2)}. \tag{6.11}$$

Taking the derivative with respect to σ_1, we obtain

$$\frac{\partial(E/u)}{\partial\sigma_1} = \frac{f_1 P_1 u_2(1 + f_2 h_2 P_2) - f_2 P_2 u_1(1 + f_1 h_1 P_1)}{[\sigma_1 u_1(1 + f_1 h_1 P_1) + (1 - \sigma_1)u_2(1 + f_2 h_2 P_2)]^2}. \tag{6.12}$$

E/u would be maximized if $\partial(E/u)/\partial\sigma_i > 0$ or

$$\frac{f_1 P_1}{(1 + f_1 h_1 P_1)u_1} > \frac{f_2 P_2}{(1 + f_2 h_2 P_2)u_2}. \tag{6.13}$$

This inequality means that the ratio of foraging rewards to risk in patch 1 exceeds that of patch 2. Again, since this inequality is free of σ_1, it follows that E/u is maximized at $\sigma_1 = 1$.

6.1.6   Linear programming to determine optimal selection of plant types

Linear programming has been used in many fields to find how choices can be made to optimize some objective function. The method, which is an alternative to the contingent models described above, was introduced to herbivore foraging by Westoby [381, 382] and Belovsky [25]. Applied to foraging, linear programming takes into account the constraints that a forager is limited by, and finds, within those constraints, what is the optimal value of a particular objective. The objective of the forager may be to maximize the amount of energy or nutrient gained during a time period, but it may also be the minimization of the time spent foraging; hence the minimization of the risk of predation.

We consider only one simple example here, described by Belovsky and Schmidt [26] for two plant species, one with chemical defenses and one without. Two constraints are considered; the limit on time to achieve certain levels of feeding on a possible mixture of the two plant types during a unit of time

$$T_{limit} \geq t_1 P_1 + t_2 P_2. \tag{6.14}$$

Here, T_{limit} is the amount of time available for consumption per unit time and P_i is the amount of biomass of plant type i that is consumed per unit time. The parameter t_i converts the quantity of food of type i, P_i, consumed per unit time into the time taken to consume it. Therefore, the sum of $t_1 P_1$ and $t_2 P_2$ must be less than T_{limit} for the herbivore to be able to consume a particular combination of P_1 and P_2. The (thick solid) line defined by the equality (6.14) in Figure 6.4 is the upper limit on amounts of P_1 and P_2 that can be taken.

The limit on the rate of digestion of a mixture of the two plant types during a unit of time

$$D_{limit} \geq d_1 P_1 + d_2 P_2. \tag{6.15}$$

The parameter d_i converts the amount of P_i consumed into what can be digested of this particular plant biomass type during that period. The sum of $d_1 P_1$ and $d_2 P_2$ must be less than D_{limit} for the herbivore to be able to take amounts of plant P_1 and P_2. The (thin solid) line defined by the equality (6.5) in Figure 6.4 is the upper limit on plant biomass taken in consideration of what can be digested in a unit of time. Any diet of the two plant types that falls below these two lines satisfies the inequalities and is thus possible. Also plotted, as a thick dashed line, is the energy constraint of the herbivore, that is, the minimum necessary energy for basal metabolism;

$$M_{limit} = e_1 P_1 + e_2 P_2 \tag{6.16}$$

where $e_1 P_1$ and $e_2 P_2$ are the amounts of energy (or nutrient, though energy will be assumed here) that can be extract from each of the plant types during a unit time. Any point along the line represents the combination of the times feeding on the two plants to obtain that same minimum amount of energy. The entire area above the minimum energy line and the constraints on feeding time and digestion capability satisfy all of the constraints.

Now suppose that this first plot of $M_{limit} = e_1 P_1 + e_2 P_2$ (thick dashed line) is for the limiting case in which the defended plant has no toxin. Any line that runs parallel to this

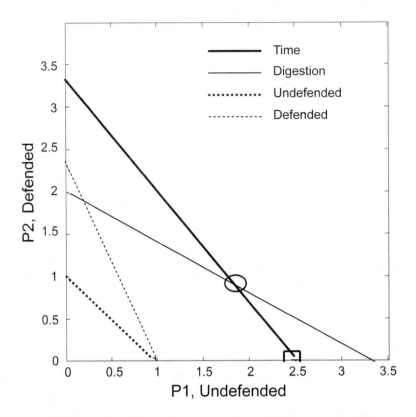

FIGURE 6.4: (P_1, P_2)-plane showing the constraints on time for feeding (thick solid line) and for digestion (thin solid line), as well as the minimum energy intake needed when plant type 2 is undefended (thick dashed line) and defended (thin dashed line). The herbivore can survive for any intake values in the area between the feeding and digestion constraints and the minimum energy intake line. The open ellipse represents the optimal selection of the two plant types when plant type 2 is undefended and the open square is the optimal plant selection when plant 2 is defended.

line represents a combination of feeding on the two plants that supplies the same amount of energy at any point along the line. Any combination of the two plants along the metabolic line (6.16) will supply the same energy intake. Imagine a line parallel to this line that is as far from the origin as possible but still has one point that is in the "possible" region defined by the time and digestion lines. This represents the maximum energy intake rate that is possible. This occurs at the open ellipse at the intersection of the time and digestion lines and shows that in this case a particular combination of the two plants maximizes the energy intake. Now assume that the defended plant produces a chemical that is a digestion inhibitor that slows down the assimilation of energy from that particular plant. This will affect the slope of the metabolism line, as more intake of the defended plant is needed to obtain the same amount of energy (thin dashed line).

There are many simplifying assumptions in the linear programming approach, such as the assumption that the constraint relationships are straight lines (see Tilman [357] for discussion of deviations from linearity). However, the approach gives some insights into how a herbivore should apportion its time for a given array of plants. Belovsky [25] showed that the predictions of the approach were in good agreement with data on three mammalian herbivores. Also, the approach can be extended beyond two plant types. Although the ability to visualize the constraints is then lost, numerical methods of linear programming can find the optimal solutions.

6.2 Snowshoe hares browsing strategy in the presence of predator

This section includes the model and analysis presented in [224] related to the dynamics of herbivores and resources on a landscape with interspersed resources and refuges.

Animals have the dual goals of obtaining resources for growth and reproduction and avoiding predation on themselves [189, 219]. Often, the goal of obtaining resources involves leaving refuges that are relatively safe from predation and foraging in areas where predation is a greater risk [35, 145]. The relative risk of predation during foraging bouts depends on environmental factors, such as how far available resources are from refuge sites. In complex landscapes, where sites with high food densities are closely interspersed with refuge sites, the risk of foraging will be smaller than in landscapes in which reaching food resources requires travelling further from the safety of predation refuges. It is therefore of interest to explore how animal foraging patterns are affected by habitat complexity and how this in turn affects aspects of the population dynamics, such as whether the population is relatively stable through time or tends to fluctuate.

To make the following analysis more concrete, we choose a particular system to analyze, the snowshoe hare (*Lepus americanus*) and its interaction with vegetation, but this also has general implications for other species, for example moose. The population dynamics of the snowshoe hare, which is a keystone species throughout the northern boreal forest of North America [200], is controlled by both predation and the supply of high-quality winter food [130, 131, 188, 200, 202, 367, 387]. Furthermore, habitat patchiness profoundly affects this tritrophic interaction [387]. The periodic irruptions of snowshoe hares known as the 10-year hare cycle only occur in landscapes that provide hares with both predation refuges in the form of dense thickets of late successional woody species such as spruce (*Picea* spp.), and interspersed patches of early successional woody vegetation such as willow (*Salix* spp.) that provide hares with an abundance of highly nutritious winter food [38, 42, 130, 131, 188, 387]. Assuming the same proportions of early and late-succession habitat, a higher degree of habitat interspersion (more mixing at a finer spatial scale) results in a higher edge/area ratio, so that hares can use sites along the edge zone between habitats to make forays into high-quality food habitat and quick dashes back to the safety of the refuge habitat when a predator threatens.

The patchy habitat that is optimal for the snowshoe hare in this forest is created largely because of frequent fires in late successional spruce forests [18, 19, 157, 243, 345, 376]. These fires, after a few years, often create irregular patches of early successional woody vegetation such as willow, aspen (*Populus tremuloides*), and birch (*Betula* spp.) [62], which provide hares with excellent food in winter [130, 131, 188, 290]. As postfire forest succession proceeds, this assemblage of highly nutritious early successional woody vegetation is replaced by dense late successional vegetation dominated by slowly growing woody species such as spruce, which are very poor winter food for snowshoe hares because they are heavily defended chemically against browsing by snowshoe hares [38, 42, 153, 227, 314], but which provide hares with refuges in which the hares can evade predators [153, 223, 387]. As Nelson et al. [266] note, "Moose and snowshoe hares often use burn edges to capture improved forage in the burns and predator protection in adjacent unburned forest." The importance of this interspersion of cover and high-quality food was emphasized by Pease et al. [290] who noted that, during the 1970–1971 hare population decline some potentially available browse was "too widely separated from suitable cover to be exploitable." The degree of interspersion of these two habitat types (early successional = good food, late successional = relief from predation) has been associated with fractal indices [165, 217, 376]. The goal of this study is to explore how the tradeoff between energy gain from foraging and safety from predation in refuges that is central to the mechanism of the 10-year hare cycle [387] might be affected by the complex nature of the landscape in which a boreal snowshoe hare population resides.

The assumption of the resource-predation tradeoffs is that the forager allocates time spent foraging to maximize fitness. A great deal of theory has investigated the effects of optimal foraging on population dynamics [2, 3, 4, 5, 6, 168, 251, 252, 331, 380], and there has been some testing of theoretical models [20, 126]. In general, both the availability of food and risk of predation have been shown to influence foraging effort (e.g., [200]). Adaptive foraging by a species in response to these factors has been shown to influence its effective functional response [2, 4, 8], such that both predator and resource levels can deviate from the standard Holling type 2 functional response [275]. The modeling approach presented here assumes that there is a negative relationship between habitat complexity; i.e., interspersion of high quality food habitat with refuge areas, and predation risk. Greater complexity reduces predation risk and therefore pushes the tradeoff in favor of increased foraging. We investigate how the influence of habitat complexity on foraging strategy affects population dynamics. We first analyze a vegetation-hare system in which predation risk is fixed at different levels. This approximates a "generalist" predator whose numbers are not influenced by the fluctuations in one particular prey species, snowshoe hares in this case. Then, we study the assumption of optimal hare foraging in an existing model of cycling in a vegetation-hare-predator model [193] in which the predator is a specialist, lynx (*Lynx canadensis*). Consideration of both models is useful because the question has often been raised whether vegetation-herbivore interactions are capable of generating population cycles in the absence of a specialist herbivore in general (e.g., [361]) and specifically in the case of the snowshoe hare (e.g., [187]).

6.2.1   Vegetation-hare-generalist predator model

Boreal forest fires create patches of foraging habitat within surrounding refuge habitat. If the "interior" of the burned areas that contain good forage are created by frequent small fires, a high degree of interspersion of foraging and refuge habitat, or high perimeter/interior ratio, is created. This means that hares do not have to travel far from areas of refuge to obtain food. They can stay along the edge between the two habitat types, making brief forays into the foraging habitat, reducing their exposure to predation. Conversely, large fires would more likely create low perimeter/interior ratios, which has the effect of reducing the total area of the burn edge habitat that contains good food adjacent to good refuge cover. The result would be an increase in the danger of foraging because hares would be forced to move farther out into the open burn areas to obtain good food. Linking foraging risk quantitatively to the fractal index of a landscape is beyond the scope of this section. In this modeling study, we will assume there is a qualitative relationship such that the risk of predation is a surrogate for landscape complexity and that predation risk per unit time spent foraging is set at a value that can be thought of as inversely proportional to the degree of interspersion.

Following King and Schaffer [193], we describe the hare-vegetation dynamics by the Rosenzweig–MacArthur (RM) predator-prey model. However, we supplement the model with the assumption, based on the above considerations, that foraging and exposure to predation are positively related, through the time spent outside of refuge sites. The predator here is assumed to be a generalist and not to vary in numbers, so the predation pressure is constant. These assumptions can be inserted into the RM model to simulate the interaction between biomass of the hare, H (kg km^{-2}) and the biomass of the hare's preferred food, poorly defended early successional woody plants, R (kg km^{-2})

$$\frac{dR}{dt} = r\left(1 - \frac{R}{K}\right)R - \frac{aRH}{1 + ahR}$$

$$\frac{dH}{dt} = \frac{\beta aRH}{1 + ahR} - dH - \mu aH. \tag{6.17}$$

Here, r (year^{-1}) is the growth rate of the vegetation, K (kg km^{-2}) is the vegetation carrying capacity, a [(kg vegetation biomass km^{-2})$^{-1}$ year^{-1}] is the foraging rate of hares on the R, and is assumed to be linearly related to the amount of time per day the hare spends foraging, h (year) is the handling time of a unit of R, β is the conversion rate from vegetation to hare biomass, and d (year^{-1}) is the natural hare mortality rate not due to predation; for example, death caused by the disease tularemia. There is also a mortality rate, μaH, that is due to predation. That rate is assumed here to be proportional to the foraging rate, a, because a is related to the time the hare is foraging. The parameter μ (loss rate of herbivore biomass per rate of vegetation biomass foraged; units of year year^{-1} kg km^{-2}) is assumed to depend inversely on the degree of interspersion, though we will not assume a specific analytic form for this relationship, and will write this simply as μ.

Assume that the hare is able to adjust the amount of time spent foraging, which is proportional to a, to whatever level of resource, R, and predation index, μ, are prevalent, in order to optimize fitness. The hare's fitness may be written as the per capita growth rate,

$$W = \frac{1}{H}\frac{dH}{dt} = \frac{\beta aR}{1 + ahR} - d - \mu a. \tag{6.18}$$

By maximizing W with respect to a, we obtain

$$\frac{dW}{da} = \frac{\beta R}{(1 + ahR)^2} - \mu = 0$$

which can be solved for a to obtain the expression for the optimal foraging rate

$$a_{\text{optimal}} = \frac{1}{hR}\left[\left(\frac{\beta R}{\mu}\right)^{1/2} - 1\right]. \tag{6.19}$$

The discussion below will focus on the case when $a_{\text{optimal}} > 0$ and

$$\left(\frac{\beta R}{\mu}\right)^{1/2} \gg 1. \tag{6.20}$$

When equation (6.19) is plugged into equations in (6.17), the new set of equations is obtained;

$$\frac{dR}{dt} = r\left(1 - \frac{R}{K}\right)R - \frac{1}{h}\left[1 - \left(\frac{\mu}{\beta R}\right)^{1/2}\right]H$$

$$\frac{dH}{dt} = \frac{\beta}{h}\left[1 - \left(\frac{\mu}{\beta R}\right)^{1/2}\right]H - dH - \frac{\mu}{hR}\left[\left(\frac{\beta R}{\mu}\right)^{1/2} - 1\right]H. \tag{6.21}$$

System (6.21) always has two boundary equilibria, $E_0 = (0, 0)$ and $E_K = (K, 0)$, and one positive equilibrium that is of biological interest (i.e., $H^* > 0$, $0 < R^* < K$ and $\beta R/\mu \gg 1$, which is denoted by $E^* = (R^*, H^*)$ where

$$R^* = \left(\frac{\sqrt{\mu\beta} + \sqrt{\mu dh}}{\beta - dh} \right)^2 = \frac{\mu}{\beta} \left(\frac{1 + \sqrt{d\mu/\beta}}{1 - dh/\beta} \right)^2$$

$$H^* = \frac{rh\mu}{\beta} \left(\frac{1 + \sqrt{d\mu/\beta}}{1 - dh/\beta} \right)^2 \left[1 - \frac{\mu}{\beta K} \left(\frac{1 + \sqrt{d\mu/\beta}}{1 - dh/\beta} \right)^2 \right] \left(1 - \frac{1 - dh/\beta}{1 + \sqrt{d\mu/\beta}} \right)^{-1}. \tag{6.22}$$

The stability analysis of E^* is conducted with the goal of determining how μ affects the stability. For ease of presentation, rewrite the system (6.21) as

$$\frac{dR}{dt} = f(R)\big(g(R) - H\big)$$

$$\frac{dH}{dt} = \beta f(R)H - dH - \alpha(R)H, \tag{6.23}$$

where

$$f(R) = \frac{1}{h}\left[1 - \left(\frac{\mu}{\beta R} \right)^{1/2} \right], \quad g(R) = \frac{rR(1 - R/K)}{f(R)}, \quad \alpha(R) = \frac{\mu}{hR}\left[\left(\frac{\beta R}{\mu} \right)^{1/2} - 1 \right].$$

The Jacobian matrix at E^* is

$$J(E^*) = \begin{pmatrix} f(R^*)g'(R^*) & -f(R^*) \\ \beta f'(R^*)H^* - \alpha'(R^*)H^* & 0 \end{pmatrix}$$

Note that

$$\det J(E^*) = f(R^*)\big[\beta f'(R^*) - \alpha'(R^*) \big] H^* = \frac{\mu}{hR^{*2}} \left(\sqrt{\frac{\beta R^*}{\mu}} - 1 \right) > 0$$

(see (6.20)) and

$$\mathrm{tr} J(E^*) = f(R^*)g'(R^*).$$

Therefore, the stability of E^* is decided by the sign of $g'(R^*)$. E^* is stable if $g'(R^*) < 0$, and it is unstable if $g'(R^*) > 0$.

The effects of μ on stability of $E^* = (R^*, H^*)$ are summarized in bifurcation diagrams (Figure 6.5a, b) based on numerical simulations of the model (e.g., Figure 6.6a, b). For large values of μ the interaction is stable, but a Hopf bifurcation occurs as μ is decreased below a threshold value. The medium range of values of μ, $\mu \approx 40$, which would correspond to landscapes of intermediate levels of interspersion of foraging areas and refuges, is where limit cycles both occur and have sizeable amplitude. The primary effect of optimal foraging in the model would be to limit the size of the limit cycles, because of the hares' cessation of feeding when vegetation R reaches the level μ/β. This is somewhat similar to the stabilizing effects of the HT3 functional response. The oscillations in Figure 6.6, in which plant biomass stays within the limits of 0.4 and 1.5×105 kg/km^2 for μ, are indicative of milder oscillations than HT2 functional responses usually produce. It should also be kept in mind that this represents edible vegetation, which is a small part of total vegetation. The equilibrium points in the extreme lower right of Figure 6.5a, b are still unstable, but the amplitude of the oscillations has decreased along with the size of the equilibrium values, which approach zero. The range of values for which there are relatively large values of plant and herbivore and limit cycles is thus relatively small.

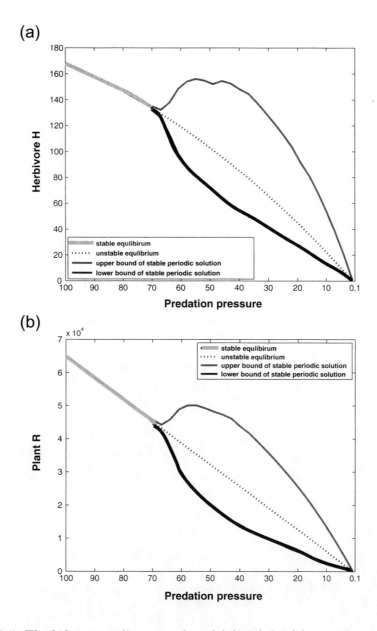

FIGURE 6.5: The bifurcation diagrams of model (6.21) for (a) vegetation and (b) herbivores. The equilibrium of the model changes with respect to parameter μ and other parameters are fixed as $r = 1.1, K = 200,000, h = 0.0014, \beta = 0.0043$, and $d = 0.5$. The solid thick-gray curve is the stable equilibrium E^*, the dotted curve is the value of the unstable equilibrium, and the solid darker curve shows the bounds of the periodic solution with the thick-darker curve showing the lower bound of the periodic solution.

6.2.2 Vegetation-hare-specialist predator model

While the above modeling of vegetation and hares allows analysis, it assumes a generalist predator community that is not variable. To take into account variation in response to hare density, we extend to the three-variable model. King and Schaffer [193] parameterized a model of snowshoe hares feeding on browse and being preyed on by a specialist predator, lynx. The model includes seasonal dynamics to incorporate scarcity of winter resources. We have adapted their model to an identical set of equations, except for converting the trophic interactions from the Monod to the HT2 form. This includes seasonality, as represented by $\Phi_s(t)$ and $\Phi_w(t)$, and effects of body condition on hare and predator mortality, δ_H and δ_P, respectively. The revised equations are

$$\frac{dR}{dt} = G\Phi_s(t)(K - R) - \Phi_w(t)\frac{aRH}{1 + ahR}$$

$$\frac{dH}{dt} = \Phi_s(t)\frac{\beta_H a_H RH}{1 + a_H h_H R} - d_H\left(1 - \frac{\delta_H \beta_H R}{1 + a_H h_H R}\right)H - \frac{(a_{H,\text{opt}}/a_H)a_P HP}{1 + a_P h_P H} \tag{6.24}$$

$$\frac{dP}{dt} = \Phi_s(t)\frac{(a_{H,\text{opt}}/a_H)\beta_P a_P HP}{1 + a_P h_P H} - d_P\left(1 - \frac{\delta_P \beta_P H}{1 + a_P h_P H}\right)P,$$

where

$$\Phi_s(t) = 1 + \epsilon_s(\Omega t), \quad \Phi_w(t) = 1 - \epsilon_w(\Omega t)$$

and $\Omega = 2\pi/\text{yr}$, $\epsilon_s = 1$ and $\epsilon_w = 0.7$. One additional difference is that we have multiplied the predation term on hares by $a_{H,\text{opt}}/a_H$, which represents the possible deviation of the predation rate when the hare forages optimally. The foraging index is variable, as it depends on both the current of vegetation and predator densities. We assume the foraging to be optimal, from which it can be shown, analogously to equation (6.19), that

$$a_{H,\text{opt}} = \frac{1}{h_H R}\left[\left(\frac{\beta(\Phi_S(t) + D_H)(1 + a_P h_P H)a_H R}{a_P P}\right)^{1/2} - 1\right]. \tag{6.25}$$

In performing simulations with this model, the goal was to vary the predation pressure on the hare population, similar to our variation of μ (Figures 6.5, 6.6), which represented the effect of the degree of interspersion of high-quality food habitat and refuge habitat on predation. Smaller values of μ represented greater ability to forage close to refuge areas.

In the present case, predation pressure was simulated by varying the predator mortality rate, d_P. Larger values of d_P had the effect of reducing predator numbers. A bifurcation analysis was performed in which d_P was varied from 0.04 to 3.0 year^{-1}, which includes the nominal value of $d_P = 1.7$ used by King and Schaffer [193]. The results show that oscillatory behavior occurs only over the lower half of this range (Figure 6.7) and that the amplitudes of the cycles are reduced from those of the King and Schaffer model. In the model with foraging optimization, a stable equilibrium occurs for $d_P > 1.4$, in contrast with King and Schaffer, for which stable equilibrium occurs only for $d_P > 3.0$. Over the range of increasing $d_P > 1.4$ in model (6.24), vegetation density decreases (Figure 6.7a), hare number density increases (Figure 6.7b), hare foraging effort increases (Figure 6.7c), and period of cycle decreases (Figure 6.7d).

As d_P decreases below 1.5, the cycling is exhibited with larger and larger amplitudes, though not as severe as in King and Schaffer until very small values of d_P are reached (Figure 6.7). The periods of the cycles increase with decreasing d_P. Only the range $0.25 < d_P < 1.4$ has cycle periods in the range observed for snowshoe hares (about 6–15 years). In the adaptive foraging model, stable oscillations occur down to values of $d_P = 0.04$, whereas in the King and Schaffer model, chaotic behavior starts to occur when d_P drops below 0.2. This stabilization may be related to the hares' relative foraging effort, $a_{H,\text{opt}}/a_H$, decreasing to levels of < 0.2 roughly during the peak of vegetation. The hare foraging effort $a_{H,\text{opt}}/a_H$ oscillates strongly for small values of d_P, though always staying below 1.0, and

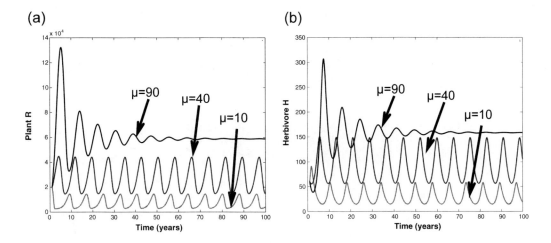

FIGURE 6.6: Time plots of (a) vegetation and (b) herbivores of model (6.21) with different μ values. Here, the other parameter values are the same as Figure 6.5. When $\mu = 10$ or 40 there are periodic solutions, whereas when $\mu = 90$, the system is stabilized. The occurrence of limit cycles of sizeable amplitude for $\mu \approx 40$ indicates that this is in the range of predation risk that might be associated with observed population oscillations of the snowshoe hare.

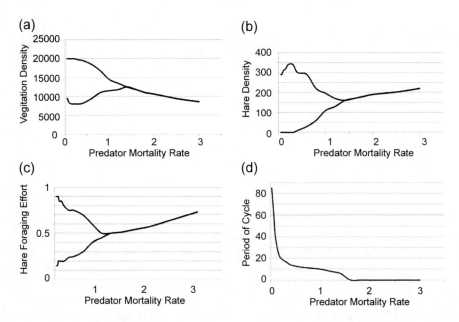

FIGURE 6.7: Results of bifurcation analysis of model (6.24) for predator mortality rate varying from 0.04 to 3.0. Plots (a)–(d) show the vegetation density R (kg/km^2), hare density H (number/km^2), foraging effort $a_{H,\text{opt}}/a_H$, and period of cycles, respectively.

the amplitude deceases rapidly as d_P increases. Both lower and upper bounds on $a_{H,\text{opt}}/a_H$ converge towards an equilibrium value of 0.5 (Figure 6.7c).

Simulation results for a particular value of d_P (1.0) show, after some initial transient behavior, the hare cycle settling to a value of about 11 years (Figure 6.8). Hare foraging effort increases during periods of high hare density and low vegetation density and is low during the peak in predator density (Figure 6.8a–d). Predator population builds up during the hare peak (Figure 6.8c).

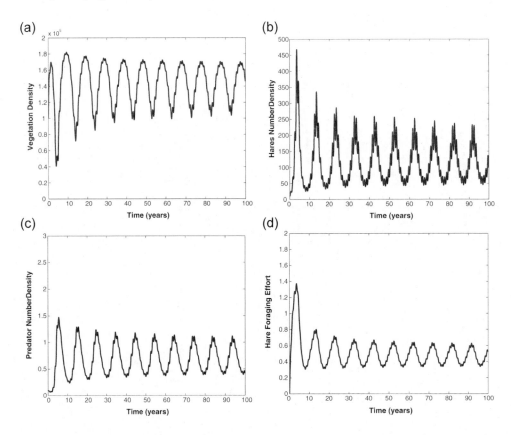

FIGURE 6.8: Dynamics for model (6.24) for $d_P = 1.0$. Plots (a)–(d) show vegetation density R (kg/km^2), hare density H (number/km^2), predator density P (number/km^2), and hare foraging effort $a_{H,\text{opt}}/a_H$, respectively.

The important result of the analysis in this section is that only for intermediate values of μ do limit cycles with large fluctuations in biomass of hares and vegetation occur. These results support Wolff's [387] hypothesis that a mosaic of patches of good food and predation refuges is a necessary condition for the existence of the 10-year hare cycle and also indicate that, if the interspersion of these two habitat types is too highly fractal, it can lead to overexploitation of vegetation by the hare, as the hare is highly protected from predation.

A main conclusion of the study of both the vegetation-hare-generalist and vegetation-hare-specialist models is that optimal foraging tended to be stabilizing over much of considered range of predation pressure, and the amplitude of oscillations tends to be decreased where they occur. In the case of the vegetation-hare-generalist model, without optimal foraging the size of the oscillations would probably lead to local extinction of the hares. This is avoided when the hares forage optimally due to the hares behaviorally adjusting foraging alternately to avoid high predation and starvation. This can be observed in the cycles shown in Figure 6.8 for the vegetation-hare specialist predator model. When vegetation is falling and predator levels have reached a low point, hare foraging increases rapidly (see [224] for more detailed explanations). When vegetation is high and predator levels are high, hare foraging falls to a minimum. Overall, the hares reduced their time spent foraging down to as low as 0.2 of the nominal value of foraging parameter, a_H, in King and Schaffer [193]. Interestingly, although the hares alternate their foraging effort along with the cycling of population numbers, the peaks of prey starvation rate and predation rate on hares occur at about the same time in the cycle. Predator number density is at a peak at this time, and prey foraging level is at a low point. Therefore, it appears that prey attempt to avoid predators by foraging less. This does dampen the effect of predation, as predator number density oscillates with changes of about 50% both above and below the mean, while the predation rate on hares deviates only about 20% from the mean. Starvation peaks, however, at the low point in the hare foraging cycle (and high point in predator density).

6.3 Glossary

Generalist predator: A generalist predator utilizes a number of different prey species.

Linear programming: This is a technique of maximizing a function in terms of a number of variables to which it is linearly related, and where there can be constraints on these variables.

Specialist predator: This predator tends to specialize in a limited number of prey species.

Chapter 7

Plant Toxins, Food Chains, and Ecosystems

In this chapter, several examples of applications of the TDFRM will be presented. The model in Section 7.1 considers the age of twig and the model is used to explain the observed cyclic snowshoe hare dynamics. Models in Sections 7.2 and 7.3 incorporate a predator of herbivore in the TDFRM with two plant species. Conditions for the invasion of more toxic plants into a static environment or an oscillatory environment are provided. The influence of predator control in vegetation composition on a boreal forest landscape is discussed. The modeling studies discussed in this chapter include results presented in [81, 95, 99].

7.1   A plant toxin mediated mechanism for the lag in snowshoe hare dynamics

This section includes some of the results presented in [81] as an application of the twig segment model (TSM) given by the equations in (4.35) in Chapter 4 to predict how hare browsing should affect twig toxicity during the hare cycle's low phase. Specifically, this explores the possibility that hare browsing in a hare peak, by reversing aging by a woody plant [258], could cause an increase in the toxicity of the Kluane snowshoe hare's preferred winter-food during the following hare low.

A necessary condition for a snowshoe hare population to cycle is reduced reproduction after the population declines. But the cause of a cyclic snowshoe hare population's reduced reproduction during the low phase of the cycle, when predator density collapses, is not completely understood. It was proposed that moderate-severe browsing by snowshoe hares upon preferred winter-foods could increase the toxicity of some of the hare's best winter-foods during the following hare low, with the result being a decline in hare nutrition that could reduce hare reproduction. A combination of modeling and experiments was used to explore this hypothesis. Using the shrub birch *Betula glandulosa* as the plant of interest, the model predicted that browsing by hares during a hare cycle peak, by increasing the toxicity *B. glandulosa* twigs during the following hare low, could cause a hare population to cycle. The model's assumptions were verified with assays of dammarane triterpenes in segments of *B. glandulosa* twigs and captive hare feeding experiments conducted in Alaska during February and March 1986. The model's predictions were tested with estimates of hare density and measurements of *B. glandulosa* twig growth made at Kluane, Yukon, from 1988–2008. The empirical tests supported the model's predictions. Thus, the conclusion is that a browsing-caused increase in twig toxicity that occurs during the hare cycle's low phase could reduce hare reproduction during the low phase of the hare cycle.

The TSM model (4.35) attempts to simulate the effects that woody plant chemical defenses may have on boreal snowshoe hare populations, which, in winter, feed almost entirely on twigs. The objective of this section is to explore the possibility that browsing by hares on their preferred winter foods during a hare peak may, by increasing the toxicity of this food, result in a reduction in hare reproduction during the following hare low. This objective can be achieved by using a combination of modeling, laboratory feeding experiments, and a long-term field monitoring experiment. The results suggest that it is likely that an increase in browse toxicity caused by hare browsing during the winter of a hare peak could reduce hare reproduction during the following hare low.

The toxicity of some of the hare's most preferred winter foods is related to the process of aging by woody plants. An excellent review of woody plant aging can be found in Chapter 4 of Kozlowski [196], which is titled "Aging." Aging is caused by the increased competition for mineral nutrients among shoots that occurs within the crown of a woody plant as the crown grows in size and increases in architectural complexity [196, 258]. Aging reduces both the lengths and the diameters of a twig's current-annual-shoots [196]. This decrease is necessarily retained during following growing seasons. This progressive annual reduction in segment size can be seen in the left stick drawing presented in Figure 7.1, which is an adaptation of the stick drawings presented in Figure 1 of Moorby and Waring [258].

When feeding upon a twig a hare bites the twig off in an OST segment (point of browsing in Figure 7.1). Then the hare consumes the twig from the point of browsing toward the tip where the twig's terminal bud is located. Initially the hare eats some toxic biomass (buds and woody internodes of YST segments). But after the hare reaches its tolerance level for the twig's toxin (or toxins) it rejects the more toxic sections of the YST and eats only the less toxic OST. The stick drawings in the middle of Figure 7.1 depict this rejected YST. An example of this foraging behavior can be seen in the photographs in Figure 3 of [307].

The twigs of some of the woody deciduous angiosperm species that the snowshoe hares prefer to eat in winter produce and store toxins in surface tissues such as the epidermis. This allocation of toxins to surface tissues prevents the toxins from killing the deeper living tissues such as the phloem and cambium that are necessary for further growth [247]. A consequence of this allocation of anti-herbivore defense toxins to surface tissues is a reduction in their concentration in a twig segment during the secondary growth that increases the twig's diameter. In most species, after the first year of shoot growth, outer tissues such as the epidermis cannot keep up with the growth in area of the twig itself [196]. Hence, sometime after the first year of secondary growth, during the formation of bark, these outer tissues die, split, and disintegrate, thus destroying any of the toxins that they contain. Moreover, since toxins are in surface tissues, increasing diameter also reduces their concentration by dilution: toxin concentration is diluted by wood production. Thus, from a snowshoe hare's perspective, the younger segments of an angiosperm twig (acronym YST) that are nearest

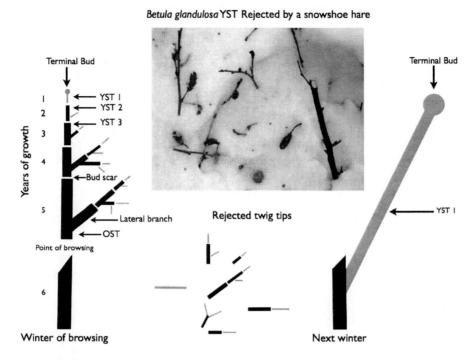

FIGURE 7.1: Stick figure drawings of a twig of an "Aged" woody plant (left drawing) and the same twig after browsing in winter by snowshoe hares has reversed aging (right drawing). The green segments are the current-annual-shoot segments (YST 1 segments). In the left stick drawing, the terminal bud scars that delineate twig segments that differ in chronological age are indicated by breaks in the twigs. The middle stick drawing represent comparatively toxic younger segments of twigs (YST segments) that have been rejected by snowshoe hares. The photograph above the stick drawings of rejected YST segments shows the YST segments of a *B. glandulosa* that have been rejected by real free-ranging snowshoe hares residing at Wiseman, Alaska.

the twig's terminal bud can be more toxic than the larger diameter older segments of the twig (acronym OST) that support the YST [308, 306]: in Figure 7.1 the toxicity of YST 1 > YST 2 > toxicity of YST 3 > toxicity of all OST, where YST1, 2, and 3 are, respectively, the current year's twig segment, last year's segment and the segment of the year before. In the left stick drawing in Figure 7.1 the terminal bud scars that mark the demarcation line between annual increments of twig growth are depicted as breaks in twigs.

From a snowshoe hare's perspective, the secondary growth driven change in twig toxicity puts the hare between a proverbial "rock and a hard spot." The rock is the comparatively high toxicity of the small diameter, nutrient rich, and highly digestible YST segments. The hard spot is the diameter at which OST segments contain so much wood that simple indigestibility and nutrient dilution renders them very poor winter-food (Figure 7.2); the larger diameter OST twig segments ultimately become starvation food. The snowshoe hare must feed in the "goldilocks zone" that contains twig segments that are both comparatively nontoxic and comparatively digestible. The snowshoe hare uses foraging behavior to solve this problem.

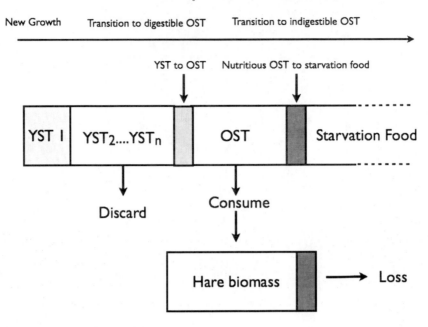

FIGURE 7.2: Conceptual diagram of how biomass is transferred among compartments in discrete time steps. On a particular time step, new growth biomass (YST 1 biomass) (light gray increment), is added to the YST compartment, biomass (medium gray increment) is transferred from YST to OST, and biomass (dark gray increment) is transferred from OST compartment to starvation food, which is twig segments that contain so much wood that they are highly indigestible and also deficient in nutrients. In addition, there is a transfer of OST biomass to hare biomass through consumption, and a loss of YST through hares discarding the comparatively high-toxicity YST segments found near the a twig's tip (in Figure 7.1 see middle stick figures and photograph).

7.1.1   Application of the model to the shrub birch *Betula glandulosa*

Numerical simulations of the TSM is carried out using parameter values suitable for the plant of interest – the shrub birch *Betula glandulosa*. This plant is selected for these reasons. 1) The *B. glandulosa* twig growth data collected at Kluane, Yukon, from 1988 to 2008 ("Shrub growth" worksheet in Kluane Monitoring Data Excel workbook [198] is sufficient to test the predictions made by the TSM simulations illustrated in Figure 7.3). 2) The twigs of *B. glandulosa*'s mature developmental phase (sensu Kozlowski [196], Chapter 3, titled "Maturation or phase change") are the Kluane hare's favorite winter-food [199]. However, at Kluane the twigs of juvenile *B. glandulosa* are rarely browsed in winter by free ranging snowshoe hares [334]. Moreover, the last two years of growth of these twigs produced a resin that strongly deterred feeding by snowshoe hares captured at Kluane: in Figure 7.1 these two segments are the YST 1 segment and the YST 2 segment. 3) This resin contained two dammarane triterpenes (papyriferic acid, acronym PA, Reichardt [304], and 3-*0*-malyonylebetulafolientriol oxide I, acronym 3*0*I, Reichardt et al. [309]). PA deters snowshoe hare feeding [307] and is also toxic to Alaskan snowshoe hares [105]. It needs to be pointed out that the discussion here should only be interpreted as an example of the potential value of the TSM as a tool to be used when studying the role anti-browsing toxic chemical defense may play in reducing snowshoe hare reproduction during the cycle's low phase. It should be pointed out that the discussion here should only be interpreted as an example of the potential value of the TSM as a tool to be used when studying the role antibrowsing toxic chemical defense may play in reducing snowshoe hare reproduction during the cycle's low phase.

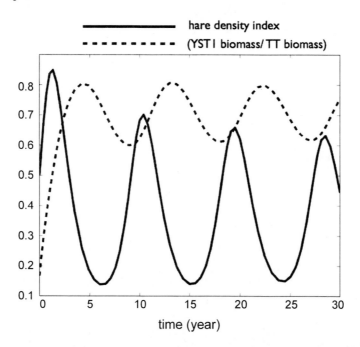

FIGURE 7.3: Results of the twig segment model (TSM) simulation. Note that the maximum magnitude of the ratio (YST biomass/TT biomass), which stands for (YST biomass/total twig biomass), occurs during the hare decline and that this ratio remains comparatively high during the subsequent low of the simulated hare cycle. The units of the hare density index and the ratio (YST biomass/TT biomass) have been normalized to range from 0 to 1. Thus, these units are dimensionless.

7.1.2 Experiments

The experimental data required to test the TSM are from the previously unpublished results of two earlier experiments. The Alaska experiment was conducted at the Institute of Arctic Biology, University of Alaska, Fairbanks, during February and March 1986. The Kluane experiment provided monitoring data from 1988 to 2008 [198].

Alaska experiment

The Alaska experiment had four objectives. The first objective was to measure the concentrations of $3\mathit{0}I$ and PA in three *B. glandulosa* twig segments: 1) the YST 1 segment of the twigs of mature phase *B. glandulosa* genets that had not been browsed for at least four years (hereafter called unbrowsed genets); 2) the OST segments of the twigs of these unbrowsed genets; and 3) the YST 1 regrowth segment produced by *B. glandulosa* genets that had been severely browsed in the previous winter (hereafter called browsed genets). The second objective was to determine if $3\mathit{0}I$, when added to an artificial diet (rolled oats) at below the concentration found in the YST 1 segment of unbrowsed *B. glandulosa* genets, deterred feeding by captive snowshoe hares. Reichardt et al. [307] had previously shown that PA strongly deterred feeding by captive snowshoe hares. The third objective was to measure the per capita daily consumption by captive snowshoe hares of the biomasses of the YST 1 segment and the older segments of unbrowsed genets. The fourth objective was the per capita daily consumption by captive snowshoe hares of the biomass of the regrowth YST 1 segment biomass that was produced by the browsed genets in the summer that followed the a browsing event (summer 1985).

Twig collection sites.

The twigs assayed for their concentrations of 3θI and PA were collected in January 1986 at each of 5 collection sites located on a 200-km-long north-to-south transect that originated in Goldstream Valley, Alaska, and ended at Denali National Park, Alaska. The twigs used to obtain the 3θI that was used in the feeding deterrence bioassay and the *B. glandulosa* twigs used in feeding trials were both collected in the Goldstream Valley site during February and March 1986.

Sampling *B. glandulosa* genets within collection sites.

Two sets of *B. glandulosa* genets were sampled within each collection site, unbrowsed genets and browsed genets. The unbrowsed genets were growing at least 100 m from the dense thickets of spruce *Picea* spp. and/or alder *Alnus* spp. that are common near Fairbanks, Alaska. The snowshoe hares in this region retreat into these thickets during a hare low in order to evade their predators [387]. Presumably distance from predation refuges explains why these genets were unbrowsed: at the time of their collection (start of a hare increase) fear of predation was probably still keeping hares very near predation refuges [153, 387]. The browsed genets were growing within 5 m of a predation refuge. In the previous winter (winter 1984–1985) hares had browsed all twigs < 5 mm diameter, hares had gnawed off most main stems at a diameter > 10 mm, and hares had also girdled (ring-barked) most main stems.

Twig collection protocol.

Twigs were collected by clipping at a diameter of 3 mm, because in the case of snowshoe hares residing in interior Alaska, large diameter twig segments are starvation food [387]. The twigs of the unbrowsed genets were separated into two parts (YST 1 segment biomass; OST segment biomass) by clipping at the terminal bud scar (diameter ≈ 1.5 mm) that separated the YST 1 segment biomass from the OST segment biomass. Since the twig segments collected from the browsed genets were 100% YST 1 biomass, they were not clipped into two parts. The resulting 3 samples of twig segments from each of the 5 collection sites were stored at $-40°$C until their concentrations of 3θI + PA were assayed in May 1986.

3θI + PA assays.

At each of the five twig collection sites, 10 unbrowsed genets and 10 browsed genets are randomly selected. Five twigs were collected from each unbrowsed genet and 5 twigs were collected from each browsed genet. After the twigs from the unbrowsed genets had been divided into two biomass types (YST 1 segment biomass, OST segment biomass), the twig segment collections were pooled by twig segment type: unbrowsed genet YST 1 segment biomass; unbrowsed genet OST segment biomass; browsed genet YST 1 regrowth biomass. This procedure resulted in 5 samples per twig segment type. In the case of the unbrowsed genets, a one-tailed paired t-test with 4 degrees of freedom $[t_{(1,4\mathrm{df})}]$ and $\alpha = 0.05$ was used for statistical analysis of the comparison of the concentrations of 3θI + PA in the YST 1 segment biomass versus the older segment biomass. Because free-ranging hares rejected the YST 1 segment biomass, this biomass is expected to have the highest concentration of 3θI + PA. A two-tailed two-sample t-test with 8 degrees of freedom $[t_{(2,8\mathrm{df})}]$ and $\alpha = 0.05$ was used to test the statistical significance of the difference in the concentrations of 3θI + PA, if any, existing between the YST 1 segment biomass of unbrowsed genets versus the regrowth YST 1 biomass of the browsed genets. The method of triterpene assay can be found in Reichardt et al. [307]. Unfortunately at 60 MHz the spectral signals used for quantifications of 3θI and PA coincide, only the combined concentration 3θI + PA is reported. Based upon relative spot size on thin layer chromatography plates (visualized by H_2SO_4 followed by heating), it is estimated that the 3θI:PA ratio to be 10:1. This result is consistent with the results obtained in [334] chemical analysis of Kluane *B. glandulosa* YST 1 segment biomass + YST 2 segment biomass.

Feeding trials with captive hares.

The twigs used for the maintenance of captive hares and the twigs used in the feeding trials with captive hares both came from the Goldstream Valley site. All twigs were collected weekly and, before use, were stored in tightly sealed plastic bags in the hare facility described by Reichardt et al. [307]. The 10 hares (6 females, 4 males; age undetermined) that were used in these feeding trials were captured November 1985 in Goldstream Valley. The hares were kept in the above-mentioned hare facility. From capture until two weeks before the feeding experiments began, the hares were fed a pure browse diet that had been previously used by the IAB to maintain captive hares for several months in winter with no weight loss and in a positive nitrogen balance: 350 grams wet mass of winter-dormant twig tips (< 3 mm diameter) from the upper crown of the mature phase (sensu Kozlowski [196]) of *B. neoalaskana* (200g), *Salix alaxensis* (100g), *Populus tremuloides* (30g), and *Picea glauca* (20g). Over the 14 days preceding the feeding trials the hares were gradually acclimated to eating *B. glandulosa* twig biomass by gradually replacing 100g of the *B. neoalaskana* twig biomass with 100g of *B. glandulosa* twig biomass that was collected from unbrowsed mature phase *B. glandulosa*. The second feeding trial measured the hares' per capita daily intake of the YST 1 regrowth biomass produced by genets that had been severely browsed in the previous winter: winter 1984–1985. The third and fourth feeding trials, respectively, tested the feeding deterrent capacity of *B. glandulosa* resin and this resin's primary constituent, 3θI. Because all of the four feeding trials used the same 10 hares, the statistical significance of all pairwise comparisons was tested with a one-tailed paired t-test with 9 degrees of freedom [$t_{(1,9\mathrm{df})}$] and $\alpha = 0.05$.

Kluane experiment

The raw data of the Kluane experiment came from the Kluane Monitoring Project Excel workbook [198]. Three worksheets were used – Shrub growth, Hares, and Lynx tracks. The data in the worksheet sheet "Shrub growth" were used to estimate aging and its reversal by hare browsing. These data are the percentage-annual-increase in the biomass of a 5 mm basal diameter *B. glandulosa* twig that is YST 1 biomass; that is, the ratio (YST 1 segment biomass/twig total biomass); in Figure 7.4 this ratio is labeled (YST 1 segment biomass/TT biomass) where TT biomass means total twig biomass. Aging was expected to reduce the magnitude of this ratio [196]. A pruning-caused reversal of aging was expected to increase the value of this ratio [258]. The sampling units used were individual 5 mm twigs collected from winter-dormant *B. glandulosa* spaced at a minimum of 15 m apart. Five millimeters is the maximum diameter at the point of browsing recorded during the comparatively high-density hare peak that occurred at Kluane in the early 1980s [335]. During the 21 years from 1988 through 2008 the number of twigs sampled per year has ranged from 129–544 twigs year^{-1} (mean ± 1 SD is 338 ± 110 twigs year^{-1}). A more detailed description of this method of measuring twig growth can be found in [255]. The density of hares can be found in the Kluane monitoring project in worksheet "Hares." Live trapping (86 traps on 20×20 grids, 30 m spacing) was used to estimate hare density. DENSITY 5 (www.otago.ac.nz/density, [94]) was used to calculate the estimate. Lynx density was estimated by the method of Sheriff et al. [330] from the number of lynx (*Lynx canadensis*) tracks counted annually in a 350 km^2 study area.

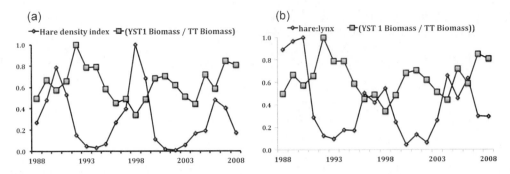

FIGURE 7.4: Results of the Kluane experiment. The highest values of the ratio (YST 1 biomass/TT biomass) occurred during the hare decline and the ratio's value remained comparatively high during the subsequent low phase of the hare cycle (Figure 7.4a): TT biomass means total twig biomass. In Figure 7.4b, the highest values of the ratio (YST 1 biomass/TT biomass) are coincident with the lowest values of the ratio of hares:lynx. The units of the ratio hare:lynx and the ratio (YST biomass/TT biomass) have been normalized to range from 0 to 1. Thus, these units have no dimensions.

7.1.3 Model predictions

The model (4.35) is applied to the plant of interest – the shrub birch *Betula glandulosa* with the assumption of a time delay $\tau_1 = 3$ years, and other parameter values given in Table 4.3. These are reasonable values for Alaskan snowshoe hares and Alaskan *B. glandulosa* as the browse species. In these simulations the limiting diameter of browsing is set to 5 mm because the data in worksheet "Shrub growth" that was mentioned above came from twigs collected by clipping at a diameter of 5 mm. The two predictions made by the simulations are represented by the line graphs shown in Figure 7.3. The important temporal pattern to note is that the magnitude of the ratio (YST biomass/twig total biomass) peaks during the simulated hare decline and remains comparatively high during the subsequent simulated hare cycle low phase. This result predicts that the toxicity of twigs, which in this case are twigs of winter-dormant *B. glandulosa* genets, will peak shortly after the hare population has peaked, and will remain comparatively high during the subsequent hare low. The increase in toxicity is the expected consequence of a browsing-caused reversal of aging that has greatly increased the percentage of a winter-dormant *B. glandulosa* twig's biomass that is composed of the most toxic YST 1 segment: compare left and right stick drawings in Figure 7.1. It is again emphasized that this predicted browsing-caused change in twig toxicity is not a result of either an induced defense (sensu Karban and Baldwin [184]) or pruning-caused reversion to the juvenile developmental phase (sensu Kozlowski [196]). It is nothing more than the browsing-caused increase in the size of the YST 1 segment that occurs when aging has been reversed by pruning [258].

Alaska experiment

The Alaska experiment's results are presented in Table 7.1. Irrespective of its source (unbrowsed genets, browsed genets), YST 1 segment biomass had about the same dry mass concentration of the potentially toxic dammarane triterpene mix $3\theta I + PA$: the concentration difference was only 0.4% dry mass. This statistically insignificant difference ($t_{(2,8df)} = 0.9409$, $p = 0.3743$) strongly suggested that the primary response to browsing in the previous winter had been a reversal of aging rather than either an induced defense (sensu Karban and Baldwin [184]) or a pruning-caused reversion to the juvenile developmental phase (sensu Kozlowski [196], Chapter 3). Either an induced defense or a reversion to the juvenile developmental phase would have caused a greater change in defense chemistry [40]. The concentration of $3\theta I$ + PA in the YST 1 segment biomass of the twigs from unbrowsed genets (4.5 $\pm$ 0.3% dry mass) was about 20 times greater than the concentration of $3\theta I$+ PA in the biomass of the older segments of twigs from unbrowsed genets (0.23 $\pm$ 0.01% dry mass). The lower concentration of $3\theta I$ + PA found in the biomass of the older segments of twigs from unbrowsed genets was caused by the domination of older segment biomass by the comparatively large diameter of the OST segments. Because of aging, the biomass of the YST segments was small in comparison to the biomass of the OST segments; in chronological time, aging progressively reduces both the lengths and diameters of YST segments [196], as depicted in the left stick drawing in Figure 7.1. Measurement of the feeding deterrent potencies of *B. glandulosa* crude resin and the purified $3\theta I$ demonstrated that both the crude resin extract and the purified $3\theta I$ strongly deterred hare feeding. The feeding trial that compared the consumption of the biomass of the YST 1 segments of the twigs from unbrowsed genets with the biomass of the twig's older segments, which were composed of the predominantly less resinous OST segment biomass, demonstrated that the hares' consumption of the older segment biomass (49.9 $\pm$ 7 g dry mass kg^{-1} hare day^{-1}) was about 50 $\times$ greater than their consumption of the YST 1 segment biomass (1.0 $\pm$ 0.4 g dry mass kg^{-1} hare day^{-1}). When the hares were only offered the YST 1 regrowth segment biomass produced by *B. glandulosa* that had been severely browsed by hares in the previous winter, the hares' per capita daily intake of this biomass was also very low (1.4 $\pm$ 0.3 g dry mass kg^{-1} hare day^{-1}), and was comparable to their consumption of the YST 1 segment biomass of the twigs from unbrowsed mature phase genets (1.0 $\pm$ 0.4 g dry mass kg^{-1} hare day^{-1}).

TABLE 7.1: Results of the Alaska experiment.

% DM 3B0I + PA	YST 1	$T_{1(4df)}$	p	OST
Unbrowsed	4.5 ± 0.3	16.0704	$p < 0.0001$	0.23 ± 0.01
Browsed YST 1	4.9 ± 0.4			
Treatment	Deterrence	$T_{1(9df)}$	p	Control
Resin	0.5 ± 0.3	10.2287	$p < 0.0001$	23.4 ± 2.5
$3\theta I$	1.6 ± 0.5	11.4572	$p < 0.0001$	21.3 ± 1.4
Intake	YST 1	$T_{1(9df)}$	p	OST
Unbrowsed	1.0 ± 0.4	6.7604	$p < 0.0001$	49.9 ± 7.3
Browsed	1.4 ± 0.3			

Notes: Unbrowsed means twigs were collected from *B. glandulosa* genets that had not been browsed for at least four years. Browsed means that twigs were collected from *B. glandulosa* genets that had been severely browsed in the previous winter (winter 1984–1985). The $3\theta I$+PA concentrations are % dry mass (% DM). Feeding deterrence (Deterrence) is grams oatmeal eaten per hare in a 6-hour period (treated oatmeal versus untreated control oatmeal). Twig segment intake (Intake) is g dry mass (DM) kg^{-1} hare day^{-1}. The data are presented as the mean $\pm$ 1 standard error of the mean. The statistical significance of the pairwise comparisons was tested with a one-tailed paired t-test. The t-values, their degrees of freedom [$T_{1(df)}$] and the probability of a Type I error (p) are shown.

Kluane experiment

At Kluane, Yukon, from 1976 through 2013 the hare cycle's period has been 8–9 years (see worksheet "Hares" in Kluane Monitoring Project Excel workbook [198]. At Kluane, from 1988 to 2008 the percentage of a 5 mm *B. glandulosa* twig's biomass that was YST 1 segment biomass peaked in the first year of each hare decline and remained comparatively high during each following hare low: In Figure 7.4 the ratio (YST 1 segment biomass/total twig biomass) biomass peaked in the first year of each hare decline and remained comparatively high during the following hare low. This result indicates that during these 21 years, as predicted by the TSM model, a reversal of aging occurred during the peak of each hare cycle. Thus, during these 21 years the presumed toxicity of twigs of the maximum diameter that Kluane snowshoe hares eat in winter should have been highest during each hare decline and following hare low. During these 21 years the ratio of hares:lynx was lowest during each hare cycle low exactly when the percentage of a 5 mm twig's biomass that was YST 1 segment biomass was comparatively high (Figure 7.4b). Since the risk of predation, and hence the chronic stress caused by fear of predation, is highest when the hare:lynx ratio is lowest [330], this result indicates that any chronic stress caused by fear of predation was highest exactly when the presumed toxicity of *B. glandulosa* twig biomass to snowshoe hares was highest.

7.1.4   Discussion

The TSM predicted that the biomass of the proportion of a *B. glandulosa* twig's biomass that is younger segment biomass (YST segment biomass) will peak in the winter of the hare decline and remain high throughout the winters of the following hare low (see Figure 7.3). The implication of this prediction is that, during the winters of the low phase of a hare cycle, hares will be forced by the comparatively high toxicity of YST segment biomass to eat the comparatively less toxic, but larger diameter, indigestible and nutrient deficient older segments of twigs that are starvation food (see Figure 7.2). The expected result of having to consume starvation food is a case of malnutrition that could reduce hare reproduction during a hare low.

The Alaska experiment demonstrated that severe snowshoe hare browsing in winter of *B. glandulosa*, like that occurring in a hare peak, does in fact, increase the lengths and the diameters of the regrowth YST 1 segment produced in the following summer. This result verified the TSM's assumption that browsing will reverse aging. This experiment further demonstrated that in winter, the YST 1 segments of *B. glandulosa*, irrespective of their diameter, have a high concentration of two feeding deterrent dammarane triterpenes, PA 30I, is also toxic to snowshoe hares [330]. This increase in toxicity may explain why, in the Alaska experiment, the captive hares' consumption of *B. glandulosa* YST 1 biomass was 50 times lower than their consumption of *B. glandulosa* older segment (OST) biomass. Increased toxicity of *B. glandulosa* YST segment biomass in general presumably explains why free ranging hares residing in interior Alaska (J. P. Bryant–Denali National Park, Goldstream Valley, Nelchina Basin; D. DeFolco–Wiseman; J. Whitman–upper Kuskokwim River Valley), residing at Kluane, Yukon (D. Hik) and residing in the Mackenzie River Valley in Canada's Northwest Territories (D. Allaire) all have selectively consumed the OST segment biomass of winter-dormant *B. glandulosa* twigs and have rejected the presumably more toxic YST biomass segments of the same twigs.

The importance of the above results is that, in concert, they experimentally demonstrate that hare browsing, by reversing plant aging, could force snowshoe hares to eat large diameter comparatively indigestible, and nutrient deficient OST biomass during the hare cycle's low phase. It is suggested that this reduction in diet quality, by causing malnutrition, could reduce hare reproduction during the hare cycle's low phase.

The Kluane field experiment verified that in the case of *B. glandulosa* that, as predicted by the TSM (see Figure 7.3), the percentage of a winter-dormant twig's biomass that is YST 1 segment biomass did peak in the winter of each hare decline and did then remain comparatively high throughout each subsequent hare low (Figure 7.4a). The importance of this result in concert with the results of the Alaska experiment is that it implies that, at Kluane, during hare declines and subsequent cyclic lows the snowshoe hares' per capita daily intake of their favorite winter-food, twigs of mature *B. glandulosa* [201], may have declined. This possibility has been supported by secondary metabolite assays of *B. glandulosa* twigs collected in winter at Kluane and by laboratory feeding trials using the resin extracted from these twigs and snowshoe hares captured at Kluane [334]. The last two years of *B. glandulosa* twig growth (YST 1 segment + YST 2 segment) did produce a resin that strongly deterred feeding by Kluane snowshoe hares. Furthermore, this resin's primary constituent was the dammarane triterpene 30I, which strongly deterred feeding by the Alaskan captive hares that were used in this study. The Kluane feeding deterrent *B. glandulosa* resin also contained the dammarane triterpene PA [334] that, in interior Alaska, strongly deters snowshoe hare feeding [307] and is also toxic to snowshoe hares [105].

Therefore, at Kluane, Yukon, browsing by snowshoe hares in the winter of a hare peak could decrease their intake of at least one preferred winter food, the twigs of the shrub birch *B. glandulosa*, during the winters of the following hare low. When the basis for the low palatability of the twigs of other winter-dormant browse species to Alaskan snowshoe hares has been chemically determined, it has almost always been found to be caused by lipophilic secondary metabolites [46, 64, 306, 307, 308] that are either potentially toxic to mammals or have been experimentally demonstrated to be toxic to mammals [43]. Thus, the TSM model may apply to more browse species than *B. glandulosa*.

A hypothesis tested by the twig segment model (TSM)

The browsing-caused reversal of aging, which increases the proportion of the twig biomass of the diameter preferred by snowshoe hares in winter that is toxic to hares, may be widespread. If this hypothesis is true, then browsing in a hare peak, by increasing the toxicity of the snowshoe hare's preferred winter-food during the following hare low, could be a cause of the low phase reduction in reproduction that is required to cause a hare population to cycle.

The other hypothesis that shows promise in explaining reduced reproduction in the hare cycle's low phase is the predator-induced chronic stress hypothesis [31, 330]. This hypothesis predicts that during the hare decline, when the hare:lynx ratio is low, fear of predation chronically stresses hares, and this chronic stress reduces reproduction during the hare decline. Furthermore, this stress is then propagated into the following hare low by maternal inheritance of high levels free cortisol.

In Figure 7.4b these two hypotheses are graphically compared by comparing the change in the ratio (YST 1 biomass/total twig biomass) of Kluane *B. glandulosa* with the change in the hare:lynx ratio at Kluane. The comparison shows that from 1988–2008, when the value of the hare:lynx ratio was low (risk of predation was high) the value of *B. glandulosa*'s (YST 1 biomass/total twig biomass) ratio was high (twig toxicity was high). This comparison therefore suggests that at Kluane the comparative importance of a predator-induced chronic stress versus a browsing-caused increase in twig toxicity in reducing hare reproduction during the cyclic needs to be determined.

In winter, the snowshoe hare generally requires a multi-species diet [30, 314], and mature *B. glandulosa* is only one of the snowshoe hare's preferred winter-foods. It is therefore suggested that the generality of the TSM model requires testing with a variety of the snowshoe hare's preferred winter-foods. Moreover, the density of hares during the hare cycle's peak has varied both in space and in time (Table 7.2). This variation implies that the intensity of browsing by hares on their preferred browse species has also varied among hare cycles. The TSM model needs testing in several locations that are characterized by a strong hare cycle, and furthermore that this testing will require the use of a long-term monitoring program such as the one at Kluane, Yukon.

TABLE 7.2:   Peak snowshoe hare densities estimated in autumn.

Region	Hares ha^{-1}	Region	Hares ha^{-1}
Northwest Territories[PC]		Alaska	
1990	2.50	Yukon-Tanana Uplands[LT]	
1999	2.38	1971	5.88
2009	1.35	Tetlin National Wildlife Refuge[PC]	
Rochester, Alberta[LT]		2008	0.95
1971	9.16	Bonanza Creek LTER Site[LT]	
Skeena British Columbia[PC]		1990	6.53
2003	1.51	1999	6.00
Kluane, Yukon[LT]		Wrangle-Saint Elias National Park[PC]	
1981	4.43	1991	2.85
1990	1.68	2001	3.93
1998	2.73	2009	2.68
2006	1.15	Kenai National Wildlife Refuge[PC]	
		1983	3.55
		1999	1.23
		2011	1.45

Notes: LT = live trapping; PC = fecal Pellet Count. Data sources: Northwest Territories (Suzanne Carrier); Alberta [188]; British Columbia (Wildlife Dynamics Consulting); Kluane Ecosystem Monitoring Project [198]; Yukon Tanana Uplands [387]; Bonanza Creek Long-Term Ecological Research Program; Tetlin National Wildlife Refuge; Kenai National Wildlife Refuge; Wrangle–Saint Elias National Park. The locations of these places can be found in the map in the Online Supplementary Materials PowerPoint named "Places."

7.2   Dynamics of the TDFRM with plant, herbivore, and predator

The model presented in this section is derived in [99], which is motivated by an attempt to explain the different food chain interaction phenomena that occur on the Tanana Flats and Yukon Flats of Alaska. In the first of these the willows are over-browsed by moose and the alders have become dominant, while in the second the willows continue to dominate. To explain this through modeling, a 4-dimensional TDFRM is formulated by introducing a predator (a wolf population) into the 3-dimensional TDFRM (4.28) discussed in Chapter 4, where the model includes two-plant species and one herbivore population. We will compare results of plant-herbivore models with and without a predator, which can help examine how the predation may have contributed to the difference in outcomes of plant-herbivore interactions between the Tanana Flats and Yukon Flats. Such models can also be used to study how the predator population may impact the succession of vegetation, and to derive invasion conditions under which a more toxic plant species can invade into an environment. These conditions provide threshold quantities for several parameters that may play a key role in the dynamics of the system. The model will be applied to a boreal ecosystem trophic chain to examine the possible cascading effects of predator-control actions when plant species differ in their levels of toxic defense.

The model (4.28) in Chapter 4 can be extended to include a predator population (e.g., a wolf population), whose density is denoted by W. Assume a Holling type 2 functional response between the herbivore and predator, i.e.,

$$f_w(H) = \frac{e_w H}{1 + h_w e_w H}.$$

The parameters e_w and h_w represent the encounter rate and handling time, respectively. Then the 4-dimensional TDFRM (which will be referred to as the 4-D TDFRM) involving two plant species, one herbivore, and one predator populations reads

$$\frac{dP_i}{dt} = r_i P_i \left(1 - \frac{P_i + c_{ij}P_j}{K_i}\right) - C_i(P_1, P_2)H, \quad i, j = 1, 2, \ i \neq j,$$

$$\frac{dH}{dt} = \sum_{i=1}^{2} B_i C_i(P_1, P_2)H - f_w(H)W - d_h H, \qquad (7.1)$$

$$\frac{dW}{dt} = B_w f_w(H)W - d_w W,$$

where

$$d_h = \mu_h + m_h, \quad d_w = \mu_w + m_w$$

with μ_i $(i = h, w)$ being the natural mortality, and m_i $(i = h, w)$ denoting the per capita death rates due to hunting (or control) of herbivore and predator, respectively. Based on similar considerations as in the simpler TDFRM, we carry out the model analysis under the following constraints on G_i:

$$\frac{1}{4h_i} < G_i < \frac{1}{h_i}, \quad i = 1, 2.$$

Here, the focus is on forest succession and issues related to the potential establishment of invasive (and more toxic) plant species. The analytical study includes bifurcation analysis to identify the invasion conditions. Without loss of generality, assume in the following analysis that plant species 1 is the resident species and plant species 2 is the invading species.

In the absence of plant 2, i.e., $P_2 = 0$, the full system (7.1) is reduced to a 3-dimensional system given below:

$$\begin{aligned}
\frac{dP_1}{dt} &= r_1 P_1 \left(1 - \frac{P_1}{K_1}\right) - C(P_1)H, \\
\frac{dH}{dt} &= B_1 C(P_1)H - f_w(H)W - d_h H, \qquad (7.2) \\
\frac{dW}{dt} &= B_w f_w(H)W - d_w W,
\end{aligned}$$

where $C(P_1) = C_1(P_1, 0)$. Consider the case in which the reduced 3-D system (7.2) is at a stable interior equilibrium. This equilibrium can be considered as a boundary equilibrium of the 4-D system (7.1) with the second plant species being absent. If this boundary equilibrium of the 4-D system is unstable due to the additional eigenvalue being positive, which suggests that solutions with initial values near the equilibrium will move away from it and enter the 4-D space, then the conditions for this instability will indicate the possibility for the second plant species to invade and establish itself. Similar arguments hold if the 3-D system has a stable limit cycle.

Let $E^* = (P_1^*, H^*, W^*)$ denote an interior equilibrium of system (7.2), i.e., E^* has positive components. Then

$$H^* = \frac{d_w}{(B_w - d_w h_w)e_w}, \quad W^* = \frac{[B_1 C(P_1^*) - d_h]H^*}{f_w(H^*)}, \qquad (7.3)$$

and P_1^* is a solution to the equation

$$g(P_1) = H^*, \qquad (7.4)$$

where H^* is given in (7.3) and $g(P_1)$ is the function defined by

$$g(P_1) = \frac{r_1(K_1 - P_1)(1 + h_1 e_1 \sigma_1 P_1)^2}{e_1 \sigma_1 K_1(1 + e_1 \sigma_1 P_1[h_1 - 1/(4G_1)])}.$$

Clearly, $H^* > 0$ if only if $B_w > d_w h_w$. The number of solutions of equation (7.4) in $(0, K_1)$ is the same as the number of intersections of the curve $g(P_1)$ with the constant line H^*, which is determined by the three values: $\frac{r_1}{e_1 \sigma_1}$, H^*, and $g(\bar{P}_1)$ where $\bar{P}_1$ is the point at which $g(P_1)$ takes its maximum. Based on the properties of the function $g(P_1)$, equation (7.4) may have 0, 1, or 2 positive solutions with $0 < P_1 < K_1$. In fact, as illustrated in Figures 7.5a and b, there is no solution if $g(\bar{P}_1) < H^*$, one solution if $H^* \leq r_1/(e_1 \sigma_1)$ (denoted by P_{1r}^*), and two solutions if $H^* > r_1/(e_1 \sigma_1)$ (denoted by P_{1l}^* and P_{1r}^* with $P_{1l}^* < P_{1r}^*$). From Figures 7.5c–e, it is clear that for each $P_{1j}^* \in (0, K_1)$, $j = r, l$, the existence of E^* depends on the sign of $C(P_{1j}^*) - d_h/B_1$ and requires that $B_w - d_w h_w > 0$.

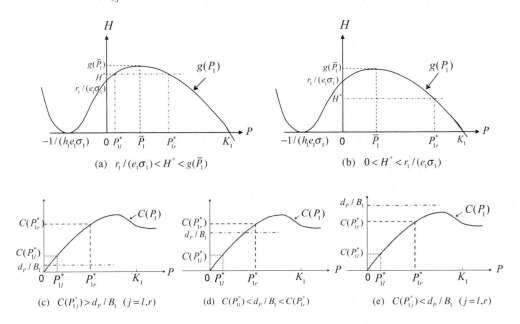

(a) $r_1/(e_1\sigma_1) < H^* < g(\bar{P}_1)$ (b) $0 < H^* < r_1/(e_1\sigma_1)$

(c) $C(P_{1j}^*) > d_p/B_1$ $(j=l,r)$ (d) $C(P_{1l}^*) < d_p/B_1 < C(P_{1r}^*)$ (e) $C(P_{1j}^*) < d_p/B_1$ $(j=l,r)$

FIGURE 7.5: Plot (a) shows two intersections of $g(P_1)$ with H^*, P_{1l}^* and P_{1r}^*, in $(0, K_1)$ when $\frac{r_1}{e_1\sigma_1} < H^* < g(\bar{P}_1)$. Plot (b) shows that there is only one intersection P_{1r}^* in $(0, K_1)$ when $H^* < \frac{r_1}{e_1\sigma_1}$. In the case when $H^* > g(\bar{P}_1)$, which is not shown in this figure, there is no intersection in $(0, K_1)$. Plots (c)–(e) are for the case corresponding to plot (a) in which both roots P_{1j}^* $(j = l, r)$ are in $(0, K_1)$. Thus, E_j^* exists if $C(P_{1j}^*) > d_h/B_1$. Therefore, the three figures (c), (d), and (e) correspond to the cases of two, one, and zero interior equilibrium points, respectively.

The existence of interior equilibria E^* is summarized in the following result.

Theorem 4 *Assume that $B_w - d_w h_w > 0$ and $P_{1l}^* < P_{1r}^* < P_{1m}$, where P_{1m} is the point where $C(P_{1m}) = \max_{0 < P_1 < K_1} C(P_1)$.*

1) If $g(\bar{P}_1) < H^*$, then system (7.2) has no interior equilibrium.

2) If $0 < H^* < r_1/(e_1\sigma_1)$, then system (7.2) has a unique interior equilibrium E_r^* when $B_1 C(P_{1r}^*) > d_h$ and no interior equilibrium when $B_1 C(P_{1r}^*) \leq d_h$.

3) If $r_1/(e_1\sigma_1) < H^* < g(\bar{P}_1)$, then

    i) both E_l^* and E_r^* exist when $B_1C(P_{1j}^*) > d_h$ $(j = l, r)$;

    ii) only E_r^* exist when $B_1C(P_{1r}^*) > d_h$ and $B_1C(P_{1l}^*) < d_h$; and

    iii) there is no interior equilibrium when $B_1C(P_{1j}^*) > d_h$ $(j = l, r)$.

For the stability analysis of an interior equilibrium $E^* = (P_1^*, H^*, W^*)$ of the reduced system (7.2), consider the Jacobian matrix at E^*

$$J(E^*) = \begin{pmatrix} C(P_1^*)g'(P_1^*) & -C(P_1^*) & 0 \\ B_1C'(P_1^*)H^* & B_1C(P_1^*) - f_w'(H^*)W^* - d_h & -f_w(H^*) \\ 0 & B_wf_w'(H^*)W^* & 0 \end{pmatrix}. \tag{7.5}$$

The characteristic equation of $J(E^*)$ has the form

$$\lambda^3 + A_1(E^*)\lambda^2 + A_2(E^*)\lambda + A_3(E^*) = 0, \tag{7.6}$$

where

$$A_1(E^*) = -\left(C(P_1^*)g'(P_1^*) + f_w(H^*)\frac{h_we_w}{1 + h_we_wH^*}W^*\right),$$

$$A_2(E^*) = C(P_1^*)g'(P_1^*)f_w(H^*)\frac{h_we_w}{1 + h_we_wH^*}W^* \tag{7.7}$$

$$+ C(P_1^*)B_1C'(P_1^*)H^* + f_w(H^*)B_wf'(H^*)W^*,$$

$$A_3(E^*) = -C(P_1^*)g'(P_1^*)f_w(H^*)B_wf_w'(H^*)W^*.$$

Because the expressions of A_i in (7.7) depend on functions of the steady state E^*, the properties of the eigenvalues or the solutions of the characteristic equation (7.6) are difficult to determine analytically. Nevertheless, it is shown in [99] that E_l^* is always unstable, and that the stability of E_r^* may switch when parameters vary, particularly when the growth rate (r_1) and toxicity (G_1) change. The nature of the stability change suggests that a bifurcation may occur. In fact, the numerical results confirm that the system has a stable limit cycle under certain conditions, as shown in Figure 7.6.

Figure 7.6 illustrates how the stability of the equilibrium E_r^* of the reduced system (7.2) depends on the growth rate r_1 (left panel) and the toxicity level G_1 (right panel), when all other parameters are fixed. It is shown in Figures 7.6a and b that the equilibrium E_r^* (only the plant component P_{1r} is shown) changes from stable (solid part of the curve) to unstable (dashed part of the curve) as r_1 increases or G_1 decreases and passes through some critical value $r_{1c} \approx 0.017$ (see (a)) or $G_{1c} \approx 65$ (see (b)). This suggests that an increased growth rate or toxicity level of the plant is likely to destabilize the system. The phase portraits for two values of r_1 and G_1 are shown in Figures 7.6c–f. It shows that for $r_1 = 0.005 < r_{1c}$ (see (c)) or $G_1 = 120 > G_{1c}$ (see (d)), the system converges to the equilibrium E_r^*. For $r_1 = 0.02 > r_{1c}$ (see (e)) or $G_1 = 50 < G_{1c}$ (see (f)), the equilibrium E_r^* is unstable and the system stabilizes at a stable periodic solution.

7.2.1   Invasion criterion of toxic plant in an equilibrial environment

In this section, the full system (7.1) is considered and conditions are presented for the invasion by a more toxic plant into an environment in which the resident plant, herbivore and predator are at either a static state or an oscillatory state of the reduced system (7.2).

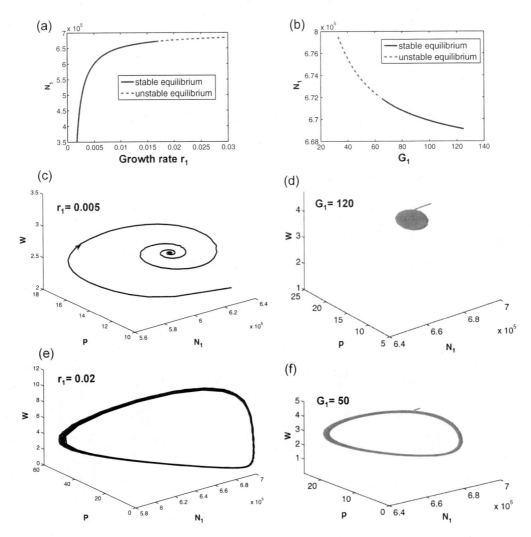

FIGURE 7.6: Bifurcation diagram and phase portraits for system (7.2) showing stability switch of E_r^* as r_1 or G_1 varies. Plot (a) shows that as r_1 increases and passes through a critical value $r_{1c} \approx 0.017$, E_r^* (only the P_{1r} component is shown) changes from stable (solid line) to unstable (dashed line). Similarly, plot (b) shows that there is a critical value $G_{1c} \approx 65$ at which E_r^* changes from unstable (dashed) to stable (solid). Plots (c)–(f) are phase portraits confirming the switch of stability of E_r^*. When E_r^* is unstable, a stable periodic orbit is present ((e) and (f), see also the text for more explanations).

Consider first the possibility of invasion in a static state. Assume that the equilibrium $E_r^* = (P_{1r}^*, H^*, W_r^*)$ of system (7.2) is asymptotically stable. Note that the interior equilibrium E_r^* of the reduced system (7.2) is a boundary equilibrium of the system (7.1) (where species 2 is absent), which is denoted by $\tilde{E}_r^* = (P_{1r}^*, 0, H^*, W_r^*)$. The stability of $\tilde{E}_r^*$ may provide conditions about whether or not plant species 2 can invade successively. In fact, the result stated in the following theorem is proved in [99]. Let

$$\lambda \doteq r_2 \left(1 - \frac{c_{21} P_{1r}^*}{K_2} \right) - H^* \frac{\partial C_2}{\partial P_2} (P_{1r}^*, 0). \tag{7.8}$$

Then the sign of this quantity λ provides an invasion criterion, as described in the following result.

Theorem 5 *Assume that the interior equilibrium E_r^* of the reduced system (7.2) exists and is stable. Let λ be defined in (7.8). Then*

(i) *The boundary equilibrium $\tilde{E}_r^*$ of the full system (7.1) is unstable if $\lambda > 0$, in which case the successful invasion by the second plant species will be expected.*

(ii) *If $\lambda < 0$, then $\tilde{E}_r^*$ becomes locally asymptotically stable, in which case the second plant species cannot invade.*

From the theoretical result stated in Theorem 5, a positive sign of λ implies the possibility for the second plant to invade into the system that is at the equilibrium $\tilde{E}_r^*$. Notice that the invasion condition $\lambda > 0$ is not expressed explicitly in terms of any particular parameter(s). This is because E_r^* was not solved explicitly. A formula for P_{1r}^* can be obtained but it does not provide clear biological insights due to its complex expression. Nonetheless, biological interpretations of the invasion condition $\lambda > 0$ might be possible from equivalent conditions to $\lambda > 0$, one of which is

$$r_2\left(1 - \frac{c_{21} P_{1r}^*}{K_2}\right) > H^* \frac{\partial C_2}{\partial P_2}(P_{1r}^*, 0).$$

The term on the left side of the above inequality represents the growth rate of plant species 2 influenced by the density of plant species 1, P_{1r}^*, the competition intensity c_{21}, and the carrying capacity K_2. The term on the right side of the inequality represents the rate of loss due to herbivore browsing, which is influenced by the herbivore density H^* and the rate of change of the consumption function C_2 determined by the density of the plant species 1 (P_{1r}^*) while the density of the invading plant species 2 (P_{2r}^*) is very low. These properties allow us to examine the roles of these factors in determining the invasion.

Although the quantity $\frac{\partial C_2}{\partial P_2}(P_{1r}^*, 0)$ does not depend on the toxicity level of invading plant species G_2 (as the partial derivative is evaluated at $P_2 = 0$), there could be an indirect effect. For example, an increased toxicity level G_2 can be associated to a decreased growth rate r_2. In addition, the relative toxicity levels of the two plant species $(G_1$ and $G_2)$ may affect the herbivore browsing preference $(\sigma_1$ and $\sigma_2)$, and the impact on the plant populations by herbivore browsing can be influenced by the density of predator (wolf). The effect of predator can be investigated by varying wolf control (reduced lifespan $1/d_w$) or by comparing the dynamics of the system without wolf (referred to as the PH system) and the system with wolf (referred to as the PHW system). Some of these cases are presented in Figure 7.7.

Figure 7.7 illustrates how the invasion threshold λ depends on the lifespan of wolf $(1/d_w)$ and the growth rate of the invading plant species (r_2) when other parameter values are fixed. Plots (a) and (b) present the bifurcation diagrams. The two curves are for the PH system (without wolf) and the PHW system (with wolf). Plot (a) shows that, in the presence of wolf, there exists a critical value $1/d_w^*$ (around 7.3 years), such that the plant species 2 cannot invade $(\lambda < 0)$ if $1/d_w < 1/d_w^*$ (dashed part of the curve) and invasion is possible $(\lambda > 0)$ if $1/d_w > 1/d_w^*$ (solid part of the curve). It also shows that in the absence of wolf, species 2 cannot invade. The difference in outcomes between the PH and PHW systems clearly illustrates that the predator population can create conditions that favor the invasion of plant species 2. This result is also confirmed via numerical simulations illustrated in the time plots (c) and (e). In (c), $1/d_w = 3 < 1/d_w^*$, in which case the plant species 2 (P_2) for both the PHW and the PH systems go to extinction. The time plot (d) is for $1/d_w = 10 > 1/d_w^*$, in which case plant species 2 (P_2) can successfully invade in the PHW

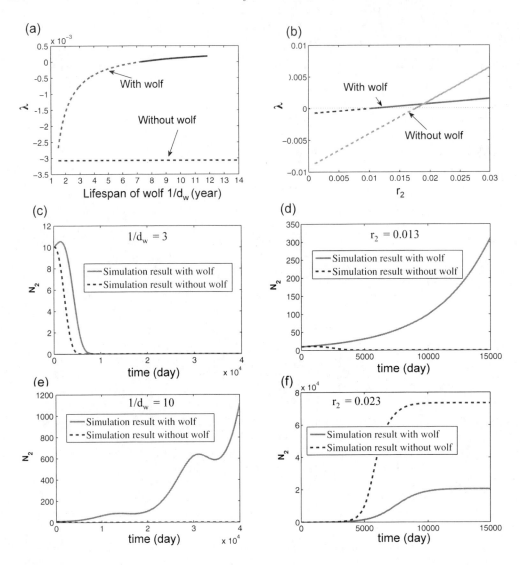

FIGURE 7.7: Plots (a) and (b) are bifurcation diagrams showing the dependence of the threshold quantity λ on the wolf lifespan $1/d_w$ and the growth rate r_2 of plant species 2 (the two curves in (b) are for the cases of with and without wolf). It shows the existence of a critical value $1/d_w = 1/d_w^*$ (see (a)) or $r_2 = r_2^*$ (see (b)) such that $\lambda > 0$ if $1/d_w > 1/d_w^*$ or $r_2 > r_2^*$, in which case invasion of species 2 is possible. Plots (c)–(f) are time plots illustrating the invasion of species 2 (P_2) for larger values of $1/d_w$ and r_2 (see the text for more explanations).

system but will die out in the PH system. Plot (b) shows that, for both the PH and the PHW systems, there exists a threshold value r_2^*, which is 0.01 for the system with wolf or 0.018 without wolf, such that $\lambda < 0$ (> 0) for $r_2 < r_2^*$ ($> r_2^*$). The time plot (d) shows that for $r_2 = 0.013$, species 2 (P_2) cannot invade without wolf (this is because $r_2 < r_2^* = 0.018$) whereas invasion of species 2 can occur in the presence of wolf (because $r_2 > r_2^* = 0.01$). In the time plot (f), $r_2 = 0.023 > r_2^*$ for both the PH and PHW system. It is observed that species 2 can invade in both systems.

7.2.2 Invasion criterion of toxic plant in an oscillatory environment

The invasion criterion is derived based on the bifurcation analysis near a limit cycle. When the stability of E_r^* switches, a stable periodic solution of system (7.2) can appear via Hopf bifurcation (identified numerically). Now consider the invasion criterion when the 4-D system (7.1) is near an oscillatory state that includes species 1, herbivore and predator populations. Let $\mathcal{C}(\bar{P}_1(t), \bar{H}(t), \bar{W}(t))$ denote a positive limit cycle (i.e., all components are positive) of the reduced system (7.2) with period $\omega > 0$. This limit cycle can be considered as an object of the full system and denoted by $\tilde{\mathcal{C}}(\bar{P}_1(t), \bar{H}(t), \bar{W}(t), 0)$. Note that for convenience the order of variables for the full system (7.1) is chosen to be (P_1, H, W, P_2).

The variational equations of system (7.1) about the limit cycle $\tilde{\mathcal{C}}(\bar{P}_1(t), \bar{H}(t), \bar{W}(t), 0)$ can be expressed as

$$
\begin{pmatrix} \dot{P}_1 \\ \dot{H} \\ \dot{W} \\ \dot{P}_2 \end{pmatrix} = \begin{pmatrix} J(\bar{P}_1(t), \bar{H}(t), \bar{W}(t)) & * \\ 0 & \phi(t) \end{pmatrix} \begin{pmatrix} P_1 \\ H \\ W \\ P_2 \end{pmatrix},
$$

where

$$
\phi(t) = r_2\Big(1 - \frac{c_{21}\bar{P}_1(t)}{K_2}\Big) - \bar{H}(t)\frac{\partial C_2}{\partial P_2}(\bar{P}_1(t), 0) \tag{7.9}
$$

and the "$*$" denotes an entry that does not affect the stability analysis. $J(\bar{P}_1(t), \bar{H}(t), \bar{W}(t))$ is the matrix corresponding to the variational equations of system (7.2) about the limit cycle $\mathcal{C}(\bar{P}_1(t), \bar{H}(t), \bar{W}(t))$, which is given by

$$
J(\bar{P}_1(t), \bar{H}(t), \bar{W}(t)) = \begin{pmatrix} a_{11} & -C_1(\bar{P}_1(t), 0) & 0 \\ B_1\bar{H}(t)\frac{\partial C_1}{\partial P_1}(\bar{P}_1(t), 0) & a_{22} & -f_w(\bar{H}(t)) \\ 0 & B_w f_w'(\bar{H}(t))\bar{W}(t) & a_{33} \end{pmatrix}
$$

with

$$
a_{11} = r_1(1 - \frac{2\bar{P}_1(t)}{K_1}) - \bar{H}(t)\frac{\partial C_1}{\partial P_1}(\bar{P}_1(t), 0),
$$

$$
a_{22} = B_1 C_1(\bar{P}_1, 0) - f_w'(\bar{H}(t))\bar{W}(t) - d_h,
$$

$$
a_{33} = B_w f_w(\bar{H}(t)) - d_w.
$$

The stability of $\tilde{\mathcal{C}}$ depends on the properties of the variational equations. Let $1, \mu_1, \mu_2$, and μ_3 be the multipliers of the Poincáre map associated with the limit cycle $\tilde{\mathcal{C}}(\bar{P}_1(t), \bar{H}(t), \bar{W}(t), 0)$ of system (7.1). Without loss of generality, assume that $1, \mu_1$, and μ_2 are the corresponding multipliers of the Poincáre map associated with the limit cycle $\mathcal{C}(\bar{P}_1(t), \bar{H}(t), \bar{W}(t))$ of the reduced system (7.2), as the subspace $\{P_2 = 0\}$ is invariant. It follows from formula (1.16) in [208] that

$$
\begin{aligned}
\mu_1\mu_2 &= \exp\Big[\int_0^\omega tr\Big(J(\bar{P}_1(t), \bar{H}(t), \bar{W}(t))\Big)dt\Big], \\
\mu_1\mu_2\mu_3 &= \exp\Big[\int_0^\omega \Big\{tr\Big(J(\bar{P}_1(t), \bar{H}(t), \bar{W}(t))\Big) + \phi(t)\Big\}dt\Big] \\
&= \exp\Big[\int_0^\omega tr\Big(J(\bar{P}_1(t), \bar{H}(t), \bar{W}(t))\Big)dt\Big] \times e^\chi,
\end{aligned} \tag{7.10}
$$

where

$$\chi = \int_0^\omega \phi(t)dt$$
$$= r_2\left(\omega - \frac{c_{21}}{K_2}\int_0^\omega \bar{P}_1(t)dt\right) - \int_0^\omega \bar{H}(t)\frac{\partial C_2}{\partial P_2}(\bar{P}_1(t),0)dt. \tag{7.11}$$

From (7.10) we have $\mu_3 = e^\chi$. The assumption about the stability of the limit cycle $\mathcal{C}(\bar{P}_1(t), \bar{H}(t), \bar{W}(t))$ implies that $|\mu_i| < 1$ $(i = 1, 2)$. Thus, the limit cycle $\tilde{\mathcal{C}}(\bar{P}_1(t), \bar{H}(t), \bar{W}(t), 0)$ of system (7.1) is locally stable if $\mu_3 < 1$ (i.e., $\chi < 0$), and it is unstable if $\mu_3 > 1$ (i.e., $\chi > 0$). This leads to the following result.

Theorem 6 *Suppose that $\mathcal{C}(\bar{P}_1(t), \bar{H}(t), \bar{W}(t))$ is a stable limit cycle of system (7.2) with period $\omega > 0$, and let χ be defined in (7.11). Then the boundary periodic orbit $\tilde{\mathcal{C}}(\bar{P}_1(t), \bar{H}(t), \bar{W}(t), 0)$ of the full system (7.1) is unstable if $\chi > 0$, and locally asymptotically stable if $\chi < 0$, in which case the second plant species can invade in the oscillatory environment.*

The condition for the sign switch of χ cannot be verified as easily as that for λ due to the lack of an explicit expression for the limit cycle $\mathcal{C}(\bar{P}_1(t), \bar{H}(t), \bar{W}(t))$ in the reduced system (7.2). Nonetheless, extensive simulations (see, for example, Figure 7.8) suggest that approximate threshold values can be determined, which help identify different cases in which plant species 2 can or cannot invade into the oscillatory environment.

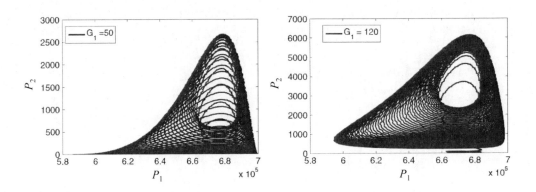

FIGURE 7.8: Numerical solutions of the full system (7.1) for two values of G_1. The figure on the left is for $G_1 = 50$, which shows that the population size of plant species 2, P_2, is going to zero as t increases. Thus, plant species 2 cannot invade. The figure on the right is for $G_1 = 120$, which shows that the population size of plant species 2 can invade ($P_2 > 0$) and the system converges to a limit cycle in the 4-D TDFRM.

Figure 7.8 demonstrates an example when the parameter G_1 is used as the bifurcation parameter. In these simulations, the parameter values used are $r_1 = 0.02, r_2 = 0.012, c_{21} = c_{12} = 0.48, d_w = 1/(8 \times 365), K_1 = 700000, K_2 = 350000, h_1 = 0.008, h_2 = 0.0001, e_1 = e_2 = 0.0001, B_1 = B_2 = 0.000051, B_w = 0.15, e_w = 0.0002, h_w = 1/0.0336, d_h = 0.00125$. Simulation results show that the reduced system (7.2) has a unique unstable positive equilibrium and a stable periodic orbit for any given $G_1 \in (32, 125)$. When plant species 2 is introduced into the system, the invasion of plant species 2 is impossible for $G_1 = 50$ (see the figure on the left). The figure on the right is for the case of $G_1 = 120$. The simulation result shows that the population size of plant species 2 (P_2) remains above zero and the solution converges to a limit cycle of the full system, indicating the invasion of plant species 2 into the oscillatory environment.

In summary, the 4-dimensional TDFRM (7.1) considered in this section is motivated by explaining the observed different patterns of plant-herbivore interactions, in which the vegetation composition is dominant by either a less toxic plant species or a more toxic plant species. The approach used here is to investigate the competition between the two plant species (e.g., willow and alder) by analyzing the threshold conditions under which the more toxic plant (alder) can invade the ecosystem, which is at a stable state (either an equilibrium or a limit cycle) with the resident (willow, which is less toxic) plant species, herbivore, and predator. Some analytical results are obtained for the reduced 3-D system (7.2) regarding the existence and stability of an interior equilibrium or a limit cycle. For example, as shown in Theorems 5 and 6, the invasion conditions for the cases of an equilibrium E_r^* and a limit cycle $\tilde{\mathcal{C}}$ are given by $\lambda > 0$ and $\chi > 0$, respectively. These threshold conditions are helpful for examining the effect of various factors (e.g., the growth rates of plant, plant toxicity levels, and the control of the predator) on the population outcomes of the plant-herbivore-predator interactions.

The analytical results are confirmed and/or extended by extensive numerical simulations of both the 4-D and the reduced 3-D systems ((7.1) and (7.2)). For example, the simulation results confirmed the existence of a stable periodic solution when the stable equilibrium E_r^* loses its stability. Moreover, these changes in stability of E_r^* are demonstrated using several key parameters including the plant growth rate r_1 and plant toxicity G_1 (see Figure 7.6). The invasion condition $\lambda > 0$ and its dependence on model parameters, as well as the role of presence of the wolf population in facilitating an invasion by the non-resident plant species are illustrated in Figure 7.7. In this particular case, ecologically speaking, the wolf population suppresses the herbivore, which prevents the herbivore from mediating "apparent competition" [160] against the invading plant. In other cases, a consequence of controlling the predator population could be an accelerated transition from less toxic to more toxic plant species in successional processes, as discussed in the next section. These results certainly demonstrated the important role that the predator may play in the succession of vegetation.

7.3 Applications of TDFRM to a boreal forest landscape

This section includes an application of the 4-dimensional TDFRM (7.1) to a boreal forest landscape presented in [95]. Field studies and theoretical modeling studies both strongly suggest that in the boreal forest of North America the chemical defenses of woody plants, via their effect on the preferences of moose (*Alces alces*) for the twigs of winter-dormant woody plants (browse), can cause spatial variation in vegetation dynamics, nutrient cycling, and biogeochemistry across landscapes [37, 43, 50, 98, 167, 190, 191, 192, 246, 256, 257, 278, 285, 288]. Furthermore, in this biome, predators such as wolves can also influence vegetation dynamics through their effect on moose density. When moose are abundant their browsing can suppress the growth of edible browse species, but when wolf predation reduces moose density, the amount of biomass of edible browse can increase because of the reduced herbivory [375]. Thus, in boreal North America, the combined effect of chemical defenses of winter-dormant woody plants and predation may reduce moose browsing and, as a consequence, may significantly affect the functioning of the ecosystem. Bridging this gap would provide opportunities for wildlife biologists, plant community ecologists, and ecosystem and regional modelers working in boreal forests to develop a common conceptual framework that allows complex interactions between trophic and successional dynamics to make informed suggestions about the management of wildlife, ecosystems, landscapes, and land-atmosphere feedbacks. Without this broad integration, management is less likely to respond effectively to interactive changes in climate and land use. In this section a mathematical model is presented that includes the effect that plant secondary metabolites, through inhibition of mammal winter browsing, may have on ecosystem processes at the landscape scale. The effects of these secondary metabolites is examined, along with predation, in the context of the Alaskan moose.

Most of the interior Alaskan boreal forest is a mosaic of patches of deciduous and evergreen trees and shrubs created by fire and the subsequent post-fire succession [62]. Fire destroys the highly flammable late successional evergreen forests that are dominated by the spruces (*Picea glauca* – white spruce; *P. mariana* – black spruce), creating the habitat required by early successional deciduous trees and shrubs. After fire, the dominant early successional deciduous species include a diverse assemblage of willows (*Salix* spp.). As post-fire succession proceeds, the vegetation progressively becomes more dominated by evergreens such as spruces, and ericaceous shrubs (e.g., *Ledum* spp.), as well as green alder (*Alnus viridis* subsp. *fruticosa*) [59].

During winter in interior Alaska, moose, which is the target species in the chapter, and snowshoe hare (*Lepus americanus*) have the same browse preferences [42, 45, 50, 244, 307], thus indicating that the same phytochemicals regulate their per capita daily intake of browse (e.g., [307]). Both moose and snowshoe hare prefer twigs of early successional species, such as the feltleaf willow (*S. alaxensis*, [45, 50, 244]), that contain low concentrations of potentially toxic secondary metabolites – phenolic glycosides [45]. Similarly, both moose and snowshoe hare eat very little twig biomass of green alder, which persists late into succession [59, 62, 244]. In fact, in the case of white and black spruces, there seems to be no report that moose even eat the twigs of these evergreen species in interior Alaska. Therefore, the relevant aspects of what is known about snowshoe hare foraging can be used with respect to chemically defended plants to moose.

For perspective on the importance of chemical defenses in this region, it should be noted that in every study of lipid-soluble secondary metabolites in boreal woody species that persist in the mid to late successional forests of interior Alaska, these species have been found to contain specific secondary metabolites that deter feeding by snowshoe hares [46, 64, 176, 254, 307, 308, 333, 347]. Furthermore, these substances have been found to be toxic to at least some mammals, including snowshoe hares [34, 43, 105, 125, 248, 249, 305]. For example, green alder [46, 64] and white spruce [56] contain two stilbenes (pinosylvin and pinosylvin-mono-methyl ether) that strongly deter feeding by snowshoe hare [64, 307, 390] and are toxic to organisms including aerobic bacteria, fungi, nematodes, insects, fish, and mammals [16, 129, 174, 220, 235, 318]. These stilbenes are broadly toxic because they are uncoupling toxins that inhibit oxidative phosphorylation in aerobic organisms [235, 318], which include moose. Both white spruce and black spruce also contain a complex mix of terpenes that are potentially toxic to Alaskan moose, because they are toxic to rumen microbes and they disrupt membranes [125, 211, 326]. Although there seems to be no experimental demonstration of the toxicity of any plant secondary metabolite to Alaskan moose or their rumen microbes, from the above it is reasonable to assume that the stilbenes pinosylvin and pinosylvin mono-methyl ether and a variety of terpenes found in green alder, white spruce and black spruce are toxic to moose and/or their rumen microbes.

The importance of plant chemical defense to the plant community is clear when the interaction of herbivores and plants is considered. A major effect of herbivores on the plant community composition is through their selective feeding on plants. This exploitation of plant biomass by herbivores is influenced by evolved plant defenses, many of which, as discussed above, are toxic to herbivores (reviewed by [341]). In response, many herbivores counteract these defenses by feeding selectively on plants to avoid accumulating the toxins (reviewed by [183]). By concentrating on plants low in toxins, the herbivore changes the composition of the vegetation toward species that are more strongly defended and thus can affect successional processes.

Nearly all terrestrial systems have a third, or carnivore, trophic level [292], that is predatory on the herbivore. When the effects of the carnivore are included in this system, it can be expected that, by exerting control on the herbivore, the predator weakens the effect of the herbivore on the plant community [295]. Therefore, predator control is likely to alleviate the effect of predators on herbivores, and so allow herbivores to exert more influence on the primary producer. In interior Alaska, predation accounts for most of the mortality of moose, and predation by wolves on moose can greatly reduce the density of moose [75, 201]. Consequently, the size of the moose population will generally be lower under low wolf control and higher under high management efforts to reduce wolf density. However, predicting accurately the effects of predators on the moose population and, therefore, also the effects of predator control on the moose, requires an understanding of the whole trophic structure, as the effects on predator mortality frequently cascade down to primary producers in boreal ecosystems [117, 201]. Thus, predator control may have not only a direct effect of lessening predation on an herbivore population, but also an indirect effect on the amount and composition of the vegetation that is food and habitat for the herbivore. Higher trophic levels can affect the processes of plant competition and, therefore, the rate and outcome of plant community succession [75, 140, 281].

7.3.1 Integration of TDFRM and ALFRESCO

The 4-dimensional TDFRM (7.1) is used in [95] to study the effect of plant toxicity and a trophic cascade on forest succession and fire patterns across a boreal landscape in central Alaska. In this application, the TDFRM is integrated into a spatially explicit simulation model, ALFRESCO, which is a spatially explicit state-and-transition model. It is a cellular automata model that stochastically simulates transitions from spruce dominated 1 km^2 spatial cells to deciduous woody vegetation based on stochastic fires, and from deciduous woody vegetation to spruce based on age of the cell with some stochastic variation. The model ALFRESCO was originally developed to spatially simulate the response of sub-arctic vegetation to changes in climate and disturbance [319, 320]. Previous research has used ALFRESCO to better understand the interactions and feedbacks among vegetation, fire, and climate [321]. ALFRESCO is stochastic in that a random number generator in conjunction with a flammability coefficient is used to ignite a fire (see Figure 7.9). The flammability coefficient is determined by the vegetation type, age of the cell, and climate [60, 89]. Climate influences fire through a weighted function of monthly average temperature for March–June and monthly total precipitation for June–July, the form of which has explained 79% of the variability in the logarithm of historical annual area burned in Alaska from 1950 to 2004 [89]. Fire spread is influenced by the flammability coefficient of vegetation and natural landscape features that are non-flammable (i.e., water, glaciers, non-vegetated areas, etc.). Burn severity is determined by a statistical relationship between fire size and burn severity obtained from remote sensing data [88]. Vegetation community composition and succession in the boreal forest are closely coupled with forest flammability, which is determined by various forest fuels [77, 146, 207]. The calibration of the model (1860 to 2007) utilizes climate data from the Potsdam Institute for Climate Impact.

ALFRESCO does not model explicitly temporally varying plant-herbivore interactions, which may have a significant impact on the succession of woody vegetation over longer time scales. By integrating the TDFRM model into ALFRESCO simulations, the dynamical interactions between plant-herbivore-predator can be taken into consideration. Particularly, TDFRM simulates woody vegetation types with different levels of toxicity, an herbivore

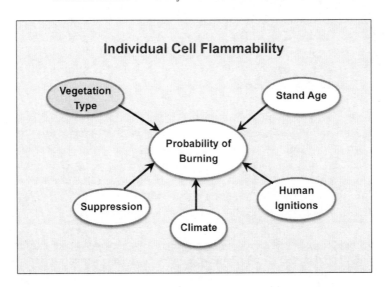

FIGURE 7.9: Stochastic factors included in ALFRESCO. The input of vegetation is assumed to be constant in the ALFRESCO simulations. The integration of TDFRM with ALFRESCO incorporates dynamic changes in Vegetation Type due to toxin-dependent herbivory and predation.

browser (moose) that can forage selectively on these types, and a carnivore (wolf) that preys on the herbivore. Here the simple succession rule in each ALFRESCO cell is replaced by plant-herbivore-carnivore dynamics from TDFRM. The central hypothesis tested in the integrated model is that the herbivore, by feeding selectively on low-toxicity deciduous woody vegetation, speeds succession toward high-toxicity evergreens, like spruce. Wolves, by keeping moose populations down, can help slow the succession. The results help confirm this hypothesis for the model calibrated to the Tanana floodplain of Alaska. The model can be used to estimate the effects of different levels of wolf control, and can be useful in estimating ecosystem impacts of wolf control and moose harvesting in central Alaska.

The example considered here is motivated by comparison of the Tanana River floodplain, where moose densities are high due to predator control and succession is prominently characterized by the competition of highly toxic (e.g., alder and spruce) species with low toxicity (e.g., willows *Salix* spp.) species, with the Yukon River floodplain, where moose densities are low due to high predation and forest succession is dominated by low-toxicity willows over a longer time period. Although there are various factors (e.g., differences in soils or climate, nitrogen pools and productivity, etc.) that contribute to these patterns, the objective here is to explore the role of plant-herbivore-predator interaction.

The TDFRM model is the same as in (7.1) except that a linear function is used for f_w with $f_w(P) = e_w P$. The TDFRM is integrated with ALFRESCO primarily by modifying the succession rule. Instead of using the cell's age to determine transition probability, the dynamics in each spatial cell is modeled, including taking into account the "toxicity" level of the cell (determined by level of coverage by spruce present in the cell), which is a dynamic variable influenced by the plant-herbivore-predator interactions (see Figure 7.10).

A key modification in the integrated model is in the rule for determining whether a cell can be classified as having undergone succession to the "spruce" state. The modified rule is based on the toxicity level of a cell, i.e., the level of spruce coverage in a cell (fraction of plant biomass in the cell that is spruce), instead of the age of the cell. In this way, the probability that a cell transitions from a "deciduous cell" to a "spruce cell" can be affected by the dynamics of the vegetation-herbivore-predator interactions. To evaluate the effect of predator control, the baseline wolf control corresponds to the case of current level of control and is used to validate the integrated modeling approach. Simulations are conducted to produce the dynamic behaviors of the two plant types and the herbivore in the TDFRM in a given cell over 100 years for the baseline scenario (Scenario 2 below). The solution trajectories exhibit a similar time scale of vegetation succession as the observed rate of boreal post-fire succession in that there is an approximately 50% reduction in deciduous density at around 50 years. For the numerical studies of the integrated models, the simulations started at year 2007, initializing the simulations with vegetation and age output maps from pre-existing ALFRESCO calibrated historical simulations (1860 to 2007). These inputs insured that our simulations start with an appropriate age/veg patch legacy dating back approximately 150 years. The MPI ECHAM5 GCM climate is used to drive the simulations from 2007 to 2099. At each time step, the dynamic changes in the plant, herbivore, and predator interactions given by the TDFRM were applied to each cell, which were used as input for ALFRESCO to determine the probability of cell transition and fire event. The outcomes from ALFRESCO, such as fire in a cell, were used as the feedback information for

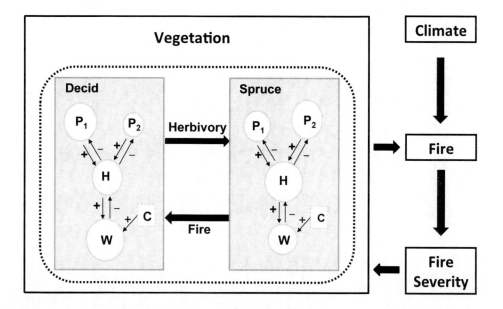

FIGURE 7.10: Conceptual diagram of the TDFRM model (inside the dashed boundary) integrated in the ALFRESCO model framework (darker). The diagram associated with TDFRM demonstrates the influences between plant species, herbivore, and predator populations. It shows two types of cells in the spatial simulations. The deciduous cell likely contains a higher density of N_1 than N_2 than does a spruce cell. Herbivory may contribute to the speed of succession, and fires will convert a spruce cell back to a deciduous cell (with various new levels of plant densities depending on the burn severity). Different sizes of circles indicate the effect of dynamic interactions between plant, herbivore (moose = P) and the predator (e.g., wolf = W). The dashed square represents an alternative food resource (e.g., caribou = C) for the predator. A "+" or "−" sign implies positive and negative impact, respectively. This diagram shows bottom-up effects from vegetation to wolves. Top-down effects can also occur, resulting from control of wolves.

the TDFRM at the next time step. The frequency of cell transitions and the burn severity of fires were recorded and used to compute the average behavior of cells for the entire area.

7.3.2   Plant toxins and trophic cascades alter fire regime and succession on a boreal forest landscape

The outcomes of the integrated models of TDFRM and ALFRESCO are presented in this section to demonstrate the impact of plant toxins and predator control on the succession on a boreal forest landscape.

To explore the possible ecological consequences under various wolf control strategies, examine three scenarios based on the following levels of wolf control:

Scenario 1. Reduced control (more wolves, fewer moose);

Scenario 2. Baseline control;

Scenario 3. Increased control (fewer wolves, more moose).

These three scenarios of wolf control correspond to per-capita wolf daily mortality (due to control) of $m_w = 0.0006$/day (hunting 20% per year), $m_w = 0.0014$/day (hunting 40% per year), and $m_w = 0.0025$/day (hunting 60% per year), respectively. The relation between m_w and the annual hunting percentage, which is denoted by q, of wolf is given by

$$m_w = -\ln(1-q)/365.$$

Changes through time in the numbers of spruce and deciduous cells of these scenarios, along with that of Scenario 2, are compared in Figure 7.11. In these simulations, the initial conditions (i.e., the initial distribution of cell age and type) use the vegetation data for the year 2007 and the time period is from 2007 to 2100.

The plot on the left of Figure 7.11 shows the ratio of spruce (the more toxic plant) to deciduous (the less toxic plant) over the time period from 2007 to 2100. For each of the three scenarios, 100 replicates were carried out. The time scenarios are represented by the thin lines and the corresponding thicker lines illustrate the mean trajectories. It can be observed that the ratio increases more rapidly in Scenario 3, followed by Scenario 2 and then Scenario 1. The plots in the right panels of Figure 7.11 show that the median transition times from deciduous to spruce for the three cases are 76 (top), 72 (middle), and 60 (bottom) years. That is, there is a difference of 16 years in the median transition age between the lowest and highest wolf control scenarios. This suggests that increased wolf control has benefited the more toxic plant species and accelerated the process of succession. This also highlights the significance of incorporating the dynamic plant-herbivore-predator interactions into ALFRESCO, especially when questions concerning vegetation succession are considered.

The interests here are not only the ratio of vegetation cover types through time, but also the spatial patterns of cover of the two vegetation types created by the dynamics in the simulations. Spatial maps of the three scenarios are shown in Figure 7.12. The top panels in Figure 7.12 present vegetation maps for the three scenarios, illustrating the differences in vegetation distributions through plots of the probability of a cell being deciduous, calculated from the dataset across all years (2007-2099) and all replicates ($n = 100$). It can be observed that map A3 for Scenario 3 (i.e., higher wolf control) shows a substantially lower (higher) density of deciduous (spruce) than Scenarios 1 and 2 (A1 and A2).

One aspect of the different spatial patterns among the three scenarios can be quantified using the spatial analysis package FRAGSTATS in the R program. It is used to analyze

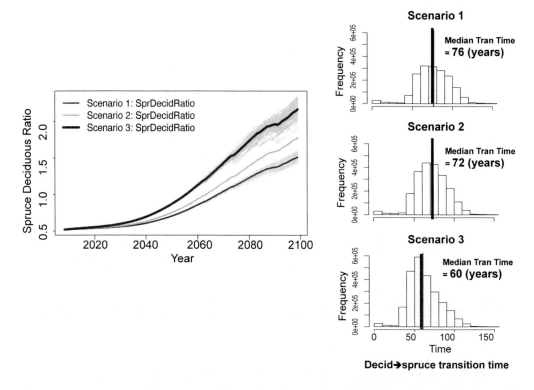

FIGURE 7.11: Simulations with 100 replicates showing the spruce to deciduous ratio for the case when the 2007 initial condition is used. The thin lines show 100 simulations for each scenario and the thicker lines show the mean trajectories for the scenarios with different levels of wolf control. The histograms on the right show the time frequencies for the transition from deciduous to spruce for three scenarios described in the Results section (Scenario 1 is low wolf control [fewer moose] and Scenario 3 is high wolf control [more moose] than the baseline scenario [Scenario 2]). The medians of the transition time (Med Trans Time) for the three scenarios are 76, 72, 60 years, respectively.

changes in moose habitat, which is defined as deciduous between 10 and 30 years post-fire [111, 237]. The numbers of deciduous 1 km² cells, which are potential moose habitat, as well as the number of distinct patches, and mean patch size, were calculated. The output shows that scenarios with high wolf control decrease the total amount of deciduous vegetation relative to scenarios with low wolf control. However, under scenarios with high wolf control, available moose habitat, which depends not just on total amount of deciduous vegetation cover, but on the amount of deciduous vegetation between 10 and 30 years of age clustered in large patches [111, 237], decreases rather than increases (Figure 7.13).

The percentage of moose habitat within the deciduous forest area (N Habitat/N Deciduous in Figure 7.13) averaged over all simulations, increased from 6.1% in the baseline case to 18.4% in the high wolf control case. The average size of deciduous habitat patches greater than 1 km² is seen to higher for high wolf control. This is illustrated in comparisons of three output figures, comparing the spatial patch distributions for low, baseline, and high wolf control for three different scenarios, representing typical minimum, average, and high numbers of deciduous spatial cells (Figure 7.13). In all three comparisons, the high wolf control scenarios exhibit greater moose habitat in the sampled year near the end of the simulation.

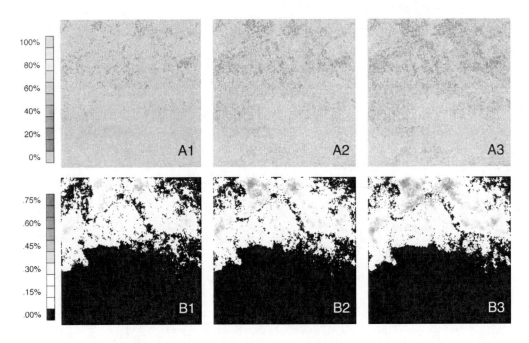

FIGURE 7.12: Results of spatial stochastic simulations of vegetation coverage three wolf control scenarios. Panels A1–A3 show the probable likelihood of a cell (1 km × 1 km) being classified as predominantly deciduous across all years (2007–2099) and replicates ($n = 100$) for 3 wolf control scenarios, in the Tanana Flats, Alaska. Scenarios correspond to three levels of wolf control: 20% (A1), 40% (A2), and 60% (A3). The 0% pixels are all Tundra. 10% likelihood of a pixel being deciduous (DE) corresponds to a 90% chance of being a spruce pixel. Panels B1–B3 show the probability (%) of a burn occurring across all years (2007–2099) and replicates ($n = 100$) for the same 3 wolf control scenarios as in A1–A3, i.e., the control levels are 20% (B1), 40% (B2), and 60% (B3).

Since both the probability that a cell will burn and the severity of the burn are higher for spruce than for deciduous vegetation, the changes in vegetation composition due to predator control may also influence fire events and the total area burned in a given region. The bottom panels in Figure 7.12 illustrate how the wolf control may affect the fire probabilities over the time period of the simulation under the same three scenarios. These maps illustrate the difference in fire distributions by plotting the probability that a cell will catch fire calculated from the same dataset used for the vegetation maps in the top panel. It is observed that the map B3 for Scenario 3 (i.e., higher wolf control) shows substantially higher fire probabilities than Scenarios 1 and 2 (B2 and B3). This is consistent with the results shown in the vegetation maps, as a higher proportion of spruce-dominated cells will be more likely to lead to larger and more severe fires.

7.3.3 More on the integration of TDFRM and ALFRESCO

For the simulation results presented in Section 7.3, the TDFRM is integrated into AL-FRESCO primarily by modifying the succession rule. As mentioned earlier, instead of using the cell's age to determine transition probability, the dynamics in each spatial cell is modeled, including taking into account the "toxicity" level of the cell (determined by the level of coverage by spruce present in the cell), which is a dynamic variable influenced by the

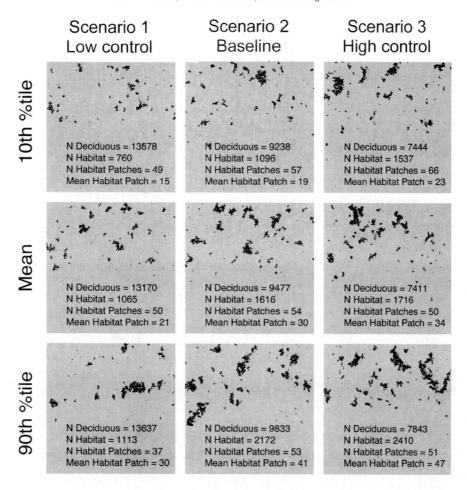

FIGURE 7.13: Simulation results of the integrated TDFRM and ALFRESCO illustrating the spatial patterns of deciduous vegetation patches generated for the three scenarios. Patches greater than 1 km^2 are shown for each of the three scenarios (low, baseline, and high wolf control). For each scenario, three specific simulation cases are shown for Year 2099; 10th percentile, mean, and 90th percentile number of deciduous spatial cells. Note that in all three cases total number of deciduous cells decreases with increasing wolf control, but both the total number of moose habitat cells (deciduous cells from 10 to 30 years since-last-fire) and Mean Habitat Patch Size (for patches > 1 km^2) increase.

plant-herbivore-predator interactions (see Figure 7.10). More detailed descriptions about the integration are included in this section.

The first step is to assign to each cell initial levels of coverage for both the more toxic and less toxic plant species, as well as for moose and wolf. This is a departure from the original ALFRESCO model, which doesn't maintain any of this detailed information for an individual cell. Each cell represents 1 km^2 of landscape, so even if a cell is classified by ALFRESCO to be deciduous, realistically it is assumed to have some small coverage of immature spruce vegetation that serves as the "seed" for eventual succession to the spruce state. With this additional information, it is possible to use the TDFRM to determine the population densities for each individual cell through time. If mean population densities are recorded on a yearly timestep from a specific solution to a TDFRM scenario, these densities can be lined up with the annual timestep that ALFRESCO uses to model a landscape. The level of spruce coverage within an initially deciduous spatial cell is then used to specify the time at which the classification of the cell switches from the deciduous state to the spruce state. The TDFRM includes the feature of being able to simulate the effects on succession of different levels of wolf control.

The integrated model assumes that the vegetation state of a cell is fixed for each step of its time unit (a year). A cell can, at a given time, have one of four different types of vegetation: upland tundra, deciduous, white spruce, and black spruce. If a cell is in the spruce state, either white or black, it will remain in that state until it burns, at which time it transitions to deciduous. A deciduous cell has a certain probability of transitioning to the respective type of spruce at each time step. In previous versions of ALFRESCO, these transitions were governed by a linear probability function of TSLF (time since last fire), the severity of the burn that preceded the current successional process, and spruce type. A number: $A_{s,b} \times \text{TSLF} + B_{s,b}$ for $s = \{$white or black spruce$\}$ and $b = \{$High, Moderate, Low burn severity$\}$ is computed annually for each deciduous pixel and tested against a random number on $(0,1)$ [178]. For ease of notation, A and B are used below for $A_{s,b}$ and $B_{s,b}$, respectively. Besides modifying how these transition probabilities are determined, no change is made for any other aspect of ALFRESCO.

A key modification in the integrated model is in the rule for determining whether a cell can be classified as having undergone succession to the "spruce" state. The modified rule is based on the toxicity level of a cell, i.e., the level of spruce coverage in a cell (fraction of plant biomass in the cell that is spruce), instead of the age of the cell. In this way, the probability that a cell transitions from a "deciduous cell" to a "spruce cell" can be affected by the dynamics of the vegetation-herbivore-predator interactions. The new rule is as follows: A number $C_{s,b} \times s_i + D_{s,b}$ is computed, where s_i is now the level of coverage of spruce in the cell of age i calculated by TDFRM. For ease of notation, C and D are used below for $C_{s,b}$ and $D_{s,b}$, respectively. A random number chosen uniformly on the interval $(0,1)$ is compared to the computed number. If the random number is smaller than the computed number, then the cell changes its state to the respective spruce category that it had before the last fire in that cell. The parameters C and D are calibrated such that the new rule (in the integrated model) represents the closest approximation possible compared to the original rule (in ALFRESCO) with the corresponding parameters A and B. A detailed mathematical description of how this is performed is provided in the following section on parameter estimation. The integration is validated by comparing the outcomes of the ALFRESCO simulations using the original (A, B) rule (i.e., in the absence of TDFRM) with the outcomes using the (C, D) rule. Once the C and D values are determined and the integration approach is validated, the influence of dynamic browsing and predation on the succession of vegetation can be examined, as presented in the previous section.

Study area.

The scenario analysis was restricted to a specific area in the boreal system. The study area is the boreal forest in interior Alaska, focusing on the Tanana Flats (GMU 20A; 17,608 km^2, of which 13,044 km^2 is moose habitat), which is located south of Fairbanks near a large urban area. Hunter access includes waterways, small airstrips, trails, and roads. However, less than 5% of the area is accessible by roads. The area is a mosaic of white spruce, black spruce, and deciduous vegetation (willow). A more detailed description of the vegetation in the Tanana Flats is documented in other literature [68]. Interior Alaska receives an annual average of 28 cm of precipitation and the annual temperature is $-3.3°$C, and ranges between -51 and $38°$C. The Tanana Flats landscape was chosen for initial testing with the integrated model due to its relatively high herbivore density and availability of ecological data.

7.3.3.1 Parameter estimation and calibration

In this section some justification is provided for the parameter values used in the simulations of the integrated TDFRM and ALFRESCO, including how the constants C and D, which are the key parameters for the integrated model, are calibrated, as well as estimates of other parameter values.

The baseline wolf control corresponds to the case of current level of control and is used to validate the integrated modeling approach. Some of the parameters are determined by simulating the TDFRM to generate the solution trajectories that exhibit a similar time scale of vegetation succession as the observed rate of boreal post-fire succession, e.g., there is an approximately 50% reduction in deciduous density at around 50 years. These parameter assumptions are further tested within the ALFRESCO modeling framework. When these within-cell dynamics from the TDFRM are incorporated into ALFRESCO with a hypothetical initial age distribution of cell ages, the outcomes for scenario 2, now in terms of numbers of cells in the "deciduous" and "spruce" states. The assumption used for the initial cell age distribution is that all cells at age zero began with the deciduous vegetation type. The comparison illustrates that the original ALFRESCO and integrated ALFRESCO/TDFRM model generate similar trajectories. This helps validate the use of our method to compare various scenarios. The constants C and D used for the computations of the modified ALFRESCO are determined using the following formulas.

First, the TDFRM is simulated for a set of values that can generate a series of values $s_i = N_2(365 * i)$, for $i = 1, \cdots, n$ for the case with browsing and predation at the "current" level, which generates the succession process similar to observed succession patterns. That is, a series is obtained from the TDFRM for the levels of spruce for each year of the simulation, where n is the final year of the TDFRM simulation. The value $n = 150$ is chosen so that all of the deciduous states in the landscape will have adequate time to transition to spruce states according to the original ALFRESCO rule.

Then, the least squares method can be used to minimize the following objective function

$$\mathcal{O} = \sum_{i=1}^{n}(s_i \times C + D - (i \times A + B))^2 \tag{7.12}$$

with respect to C and D. By solving the equations

$$\frac{\partial \mathcal{O}}{\partial C} = 0, \quad \frac{\partial \mathcal{O}}{\partial D} = 0$$

the following C and D can be obtained

$$C = \frac{A}{\sigma_s^2}\left(\frac{\sum_1^n is_i}{n} - \frac{(n+1)\bar{s}}{2}\right),$$

$$D = \frac{A(n+1)}{2} + B - \frac{A\bar{s}}{\sigma_s^2}\left(\frac{\sum_1^n is_i}{n} - \frac{(n+1)\bar{s}}{2}\right),$$

where

$$\bar{s} = \frac{\sum_1^n s_i}{n},$$

and

$$\sigma_s^2 = \frac{\sum_1^n (s_i - \bar{s})^2}{n}.$$

One can check that this actually constitutes a minimum for the function in (7.12) by verifying the following conditions:

$$\frac{\partial^2 \mathcal{O}}{\partial D^2} = 2n > 0,$$

and

$$\begin{vmatrix} \dfrac{\partial^2 \mathcal{O}}{\partial C^2} & \dfrac{\partial^2 \mathcal{O}}{\partial CD} \\[2mm] \dfrac{\partial^2 \mathcal{O}}{\partial DC} & \dfrac{\partial^2 \mathcal{O}}{\partial D^2} \end{vmatrix} = 4n^2\sigma_s^2 > 0.$$

Other parameter values are chosen based on the following justifications.

Carrying capacities K_1 and K_2.

It is assumed that an overall plant biomass carrying capacity is characteristic of boreal forests. According to [32], ovendry biomass density in the North American boreal forest ranges from roughly 1 to 10 kg m^{-2}. We choose values in the lower end of the range of values for the region studied here, so that total carrying capacity of all woody species is assumed to have values in the range of 1,000,000 to 2,000,000 kg km^{-2}.

Intrinsic growth rates r_1 and r_2.

Intrinsic growth rates of deciduous species and spruce differ significantly on the same sites. Data on growth of seedlings (see [61]) is used to estimate these rates for the model. The annual intrinsic growth rate for the deciduous types was assumed to be 0.10 and that of spruce to be 0.025. The daily rates were computed by dividing each of these rates by 365 to obtain 0.00027 and 0.000068 day^{-1}.

Moose handling time of forage h_1 and h_2.

To obtain the handling time per unit resource, it is needed to find the maximum rate of biomass intake per unit moose biomass. It was found [324] that during the winter a 347-kg moose, fed on willow, ingested 52.4 g per day per body weight raised to the 0.75 power; i.e., BW$^{0.75}$. This means that the daily intake of the moose was about $52.4 \times 347^{0.75} = 4.2$ kg. Given that summer ingestion rate may be twice this value, choose a net daily intake of the 347-kg moose to be 6 kg, or a fraction 0.017. The inverse of this is the handling time, or, roughly, $h_1 = 60$ days. It is assumed that the handling time for the spruce is larger, $h_2 = 100$ days.

Conversion fraction for moose B_1 and B_2.

The conversion factor from browse to moose biomass is assumed to be the same as the metabolizable energy estimated by [324] to be 0.278. This is the efficiency of gross production. Assuming that about half of the assimilated energy is lost to respiration, then the net production efficiency will be $B_1 = B_2 = 0.14$.

Conversion fraction for wolf, B_w.

The conversion fraction of moose to wolf biomass depends on both the fraction of moose biomass that can be ingested and the efficiency of wolves in assimilating the amount ingested. The fraction of moose carcass consumed has been estimated as 0.75 [23]. Due to the loss in the assimilation process of ingested biomass, the conversion ratio is estimated to drop to 0.7. Assuming that 0.80 of the gross productivity is lost to respiration, then the conversion ratio is $B_w = 0.5 \times 0.7 = 0.35$.

Estimation of the remaining TDFRM parameters in the integrated model is based on observed rates of post-fire secondary succession in the boreal ecosystem. As a starting point, assuming that on average, across both spruce species and multiple potential trajectories resulting from variations in burn severity, it takes around 100 years for a deciduous patch to transition to a spruce patch [87, 363] This is consistent with ALFRESCO's original calibrated TSLF successional parameterizations ($A_{s,b}, B_{s,b}$), which allow deciduous to spruce transitions ranging from 30 to 180 years post-fire across the spruce species and burn severity scenarios. Given the parameter values specified, the TDFRM (7.1) is fitted to the typical trajectory to determine the values for the remaining parameters (which were not easily obtained from the literature or other sources) including c_{12}, c_{21}, e_1, e_2, e_w, and r_w. The simulations demonstrate a fit under the assumption that the typical trajectory reflects an average level of browsing and predation (e.g., a hunting rate of 20% per year for moose and wolf corresponds to an extra mortality of $d_h = -\ln(0.8)/365 = 0.0006$ per day and $d_w = -\ln(0.6)/365 = 0.0014$ per day). The fitted parameter values are $e_1 = 5 \times 10^{-7}, e_2 = 1 \times 10^{-8}, c_{12} = 0.6, c_{21} = 0.9, e_w = 6.5 \times 10^{-7}, r_w = 0.0007$. The values of G_i (the toxicity levels of plant species i for $i = 1, 2$) are obtained from h_i under the constraints $G_2 < G_1$ (i.e., plant species 2 is more toxic than plant species 1) and $1/(4h_i) < G_i < 1/h_i$ (so that the maximum consumption rate C_i of plant species i is G_i and the function $C_i(N_1, N_2)$ remains nonnegative). The values $h_1 = 60$ and $h_2 = 100$ lead to $G_1 = 0.02$ and $G_2 = 0.005$ in order to make toxin effects determine the saturation of herbivore consumption rates. Assuming that the mean lifespan of a moose is nine years and the mean lifespan for a wolf is ten years, then $\mu_h = 0.0003$ and $\mu_w = 0.00027$. The units for population densities for all variables have been chosen in a way that makes conversion between energy unit and number densities of moose and wolves. (For example, one moose is equivalent to 350 units of energy while a wolf is equivalent to 40.) In this case, $B_1 = B_2 = 0.14$, and $B_w = 0.35$.

A summary of the parameter values, units, and sources is listed in Table 7.3.

TABLE 7.3: Definition of parameters used in simulations. Sources for some of the parameters are provided and the rest of the parameters were determined by calibration simulations.

Par	Definition	Value	Unit	Source
r_i	Plant intrinsic growth rate	$(0.0006, 0.03)$	day^{-1}	[61]
c_{12}, c_{21}	Competition coefficient	$(0.2, 2)$		
K_i	Carrying capacity of plant species i	$(3 \times 10^5, 7 \times 10^5)$	kg	[32]
G_i	Maximum unit of plant species i a unit of moose can consume per day	$(0.005, 0.02)$	kg	
h_i	Time for handling one unit of plant species i per unit of herbivore	$(50, 100)$	day	[324]
e_i	Rate of encounter per unit of plant species i	$(10^{-8}, 5 \times 10^{-7})$		
σ_i	Fraction of effort applied to foraging for plant species i	$\sigma_1 + \sigma_2 = 1$		
B_i	Conversion constant (moose biomass per unit plant species i)	0.14		[324]
B_w	Conversion constant (moose to wolf in biomass)	0.35		[23]
e_w	Encounter rate per unit of wolf per unit of moose	6.5×10^{-7}		
r_w	Additional growth rate of wolf	0.0007		
μ_h	Natural death rate of moose	9	year^{-1}	
μ_w	Natural death rate of wolf	10	year^{-1}	
d_h	Death rate of moose due to hunting	0.0006		
d_w	Removal rate of wolf due to control	Varied		
$N_i(0)$	Initial level of plant i one year after a fire			
$H(0)$	Moose initial level one year after a fire			
$W(0)$	Wolf initial level one year after a fire			

Notes: $i = 1, 2$ for plant species i. All parameters are given for 1 km^2. The interval (a, b) means a range of values from a to b.

7.3.4   Discussion

In this application, by integrating the TDFRM with ALFRESCO we demonstrated that dynamic plant-herbivore-predator interactions can have a substantial impact on vegetation succession in the boreal ecosystem. This is due to an apparent trophic cascade from the predator to the herbivore and then to vegetation. Because the herbivore in the boreal ecosystem modeled here, moose, feeds selectively on lower-toxicity vegetation, higher herbivore density promotes succession toward higher-toxicity vegetation and lower herbivore density tends to reduce the rate of succession. Thus the trophic cascade affects the vegetation community composition and succession. These results show that management of predators due to both harvest regulations and predator control produces noticeable differences on vegetation composition and succession. The results from this research can be used to manipulate predator management to optimize the balance between plant species with different toxicity levels and herbivore (e.g., moose) densities.

The modeling study was applied to a specific ecosystem, the Tanana Flats of Alaska. By applying the integrated model to three different wolf control scenarios, it illustrated how various levels of wolf control may affect the time scale of succession in the Tanana Flats. The model outcomes indicate that higher levels of wolf control can lead to an accelerated transition of land cover from deciduous to spruce. These findings provide a possible explanation for the observed pattern in Tanana Flats, where, along a gradient of higher moose densities and higher moose:wolf ratios, more toxic plants tend to replace less toxic plants over successional time. Thus the model is consistent with empirical findings [50] and suggests an important role of herbivory in the successional process. The comparison between ALFRESCO simulations with and without incorporating TDFRM demonstrated the advantage of including dynamic changes of plant, herbivore, and predator populations in ALFRESCO. The TDFRM allows inclusion of the effects of trophic structure on succession.

As was mentioned earlier, this study was "motivated" by the differences between the Tanana floodplain, where there has been extensive predator control and low predator densities, and the Yukon floodplain, where there has been less predator control and predator densities are high. The simulations presented in this section have used the model calibrated to the Tanana floodplain, and the results are consistent with the observed patterns in both areas. For example, Figure 7.11 shows that increasing predator control, as found in the Tanana floodplain compared with the Yukon floodplain, may lead to an increased spruce to deciduous ratio (left) and an accelerated succession process (right). It also demonstrates the joint effects of toxin-determined functional response for herbivore browsing and the interactions between plant-herbivore-predator have significant implications for the dynamics and management of boreal ecosystems. For example, the predator control that characterizes wildlife management in many parts of Alaska may increase the density of moose and provide more hunting opportunities, as intended. It may, however, also speed the succession from deciduous to spruce forest and increase forest flammability and fire risk near communities where this predator control is concentrated, leading to unintended negative societal consequences. In addition, due to the lower albedo of spruce than deciduous species, more rapid transition to spruce would act as a positive feedback to regional warming, augmenting the vegetation-induced increase in forest flammability [21].

The analysis in this section also indicated the effect of top-down effects on the spatial pattern of moose habitat. In particular, the analysis of moose habitat resulted in the counterintuitive result that under high wolf control, the type of habitat that was highly attractive to moose, large patches of deciduous vegetation between 10 and 30 years of age, increased, even though the area of deciduous cover decreased. This result was not an artifact of a single year being chosen to calculate results (Year 2099 in Figure 7.13), because the differences in the trend in the patch size of moose habitat among the three scenarios remained constant over our study period (results not shown here). Therefore, it is believed that the higher amount of moose habitat in large patches between 10 and 30 years is a robust result. This counter-intuitive result concerning available moose habitat is likely due to the higher frequency of large fires that occurs when the succession toward spruce dominance is faster. Larger and more severe fires increase the likelihood of larger contiguous age cohort patches. Because the total amount of deciduous vegetation decreases, but desirable moose habitat increases under high wolf control, the percentage of the deciduous forest that is highly attractive to moose increases. There are some possible negative consequences of this increase in the ratio of high-quality to lower-quality habitat across the landscape mosaic, as it could result in over browsing of the excellent deciduous patches. Fewer deciduous trees in the high wolf control simulation are over 30 years of age, an age at which they are likely to be above the browse height of moose, and at which they are a secure source of seedlings for future generations [177]. When moose consume poorer quality food they reduce the rate of consumption due to the increased time of digestion [257]. Thus a mixture of habitat areas of different quality helps to regulate browsing on the landscape and may help stabilize the population. Furthermore, wolf control also has the direct effect of an increased number of moose, which increases the likelihood of overbrowsing. An increase in the number of moose could negate the benefits of fire due to increased browsing pressure.

7.4 Glossary

4-D TDFRM: 4-dimensional toxin-determined functional response model or 4-dimensional TDFRM.

Hare cycle low phase: This refers to the phase of the snowshoe hare-lynx cycle when both the hare and lynx have reached very low numbers. It has long been a puzzle that hare recovery during this phase is slow, although their lynx predators are extremely rare and vegetation is relatively abundant.

Invasion criterion: In using mathematical models to study the possibility that a population can invade an ecological community, the differential equation for the invading species is examined, assuming the invading population is at very low levels and the resident community is on its attractor, which can be either an equilibrium point or an oscillatory state (e.g., a limit cycle).

Reversal of aging in plants: After heavy browsing, due to decreased competition within a plant for nutrients, many mature woody plants can produce shoots that are characteristic of the juvenile phase of the plant and larger than those of the mature plant and more likely to have mechanical or chemical defenses. These defenses differ from induced defenses, as they are similar to those of unbrowsed younger woody plants.

Spatially explicit modeling: Spatially explicit models in ecology model the dynamics and spatial movement of ecological populations or communities on landscapes that take into account the spatial features and configurations of actual landscapes.

Twig segment model (TSM): This model is based on the idea of reversal of aging in some woody plants. It takes into account that the tips of new twigs after heavy browsing are richer not only in nutrients, but also toxins, so that there is a gradient from greater to lesser concentrations of both nutrients and toxins, going away from the new segments of twigs toward the older segments.

Trophic cascade: Trophic cascades in tritrophic (three-level) food chains occur when the species at the trophic level at the top of the chain, the carnivore trophic level, suppresses the trophic level below it, usually the herbivore trophic level. This in turn releases the autotroph or plant trophic level from limitation by herbivory. In a four-level trophic chain, the effects may also cascade down, but in this case second-order carnivores limit first-order carnivores, so that herbivores are not limited and autotrophs are.

Chapter 8

Fire, Herbivory, Tree Chemical Defense, and Spatial Patterns in the Boreal Forest

8.1   Introduction

We have described the modeling of interactions of herbivores with plants with chemical defenses. We have noted that the browsing responses of herbivores to differential toxicity in plants can have effects on plant communities. In Chapter 7, for example, it was shown that the interaction of disturbances, such as fire, with herbivory, affects spatial patterns of deciduous and coniferous trees in boreal forests. Here we discuss the broader implications of herbivore-plant interactions on biogeographical patterns of two species of birch with respect to their chemical defenses.

An important biogeographical question is: What ecological factors have affected the distributions of tree species in boreal North America since the last glacial maximum (LGM, [385])? In the case of *some* species a previously unrecognized factor affecting their distribution since the LGM may have been an interaction between forest fire and toxin-dependent selective browsing in winter by mammals (hereafter selective browsing) such as the snowshoe hare *Lepus americanus* and the moose *Alces alces*. Our objective in this chapter is to show how the implications of the mathematical modeling presented in this book can be used to theoretically explore this specific hypothesis and other similar macroecological hypotheses.

8.2   Fire's relationship to browsing mammal abundance

It is important first to understand the important role of fire in the boreal forest. In boreal North America, fire is essential to the existence of most browsing mammal populations [130]; references in Fox 1978 [107], and especially snowshoe hare populations [131]. This is because fire creates most of the habitat mosaic required by these herbivores. Fire destroys late successional evergreen forests that are dominated by the slowly growing spruces *Picea mariana* and *P. glauca* that are very effectively defended against browsing by resins [38, 42] that contain toxic lipid-soluble secondary metabolites such as the monoterpene camphor that deters feeding by snowshoe hares [333]. However, even though spruce is a comparatively poor winter-food for boreal browsers [42], dense spruce thickets do provide the protective cover that these herbivores use to evade their predators [153, 387]. The recently burned patches within an unburned spruce forest matrix are colonized by the more rapidly growing and comparatively poorly defended early successional deciduous species such as the willows (*Salix spp.*), quaking aspen (*Populus tremuloides*), and the birches that are the preferred winter-foods of most North American boreal browsing mammals [38, 42]. Thus, in boreal North America most of the optimal habitat for browsing mammals occurs at the edge of burns where browsers have ready access to both good predator escape cover (spruce forest) and good food (fast growing deciduous species in recently burned patches). For this reason in boreal North America the abundance of browsing mammals is often greatest in a landscape that contains patches of poorly defended early successional deciduous woody species within a late successional spruce matrix ([266, 340] and references therein). And where the abundance of these browsing mammals is greatest, the intensity of their selective browsing on the recruitment of rapidly growing deciduous tree species such as *B. neoalaskana* and *B. papyrifera* is generally most intense.

8.3 Antibrowsing chemical defenses of *B. neoalaskana* and *B. papyrifera*: Benefit and cost

We will see that this positive effect of fire on herbivory may have had a further effect on the geographic distributions of two widespread North American tree birches, the diploid (2 n = 28) Alaskan resin birch *Betula neoalaskana* Sargent (previously named *B. resinifera*, [90]) and the polyploid (2 n = 56, 70, 84) white birch *B. papyrifera* Marshall (see upper 2 maps in Figure 8.1). The juvenile developmental phase (seedlings, saplings [196, 293]) are both preferred winter-foods of boreal browsers such as snowshoe hare and moose ([30, 203, 288]). But feeding trials using snowshoe hares [44] and moose browsing surveys done in birch provenance plantations (M. Carlson per. comm.) have consistently shown that when selecting between saplings of these two birch species that *B. papyrifera* is highly preferred over *B. neoalaskana*.

The primary taxonomic trait used to distinguish juvenile *B. neoalaskana* from juvenile *B. papyrifera* is the presence of resin glands on twig tips: tips of juvenile *B. neoalaskana* twigs are densely covered with resin glands [90, 272]; tips of juvenile *B. papyrifera* twigs contain few if any resin glands, and any glands that do occur are small and can, therefore, produce little resin [90, 273].

Juvenile *B. neoalaskana*'s twig-tip resin glands produce the dammarane triterpene papyriferic acid (PA, [304]). Feeding experiments have demonstrated that PA strongly deters snowshoe hare feeding [307] and toxicological experiments have demonstrated that PA is toxic to a variety of mammalian herbivores [249] including the snowshoe hare [105]. Using wapiti (*Cervus canadensis*) rumen fluid, Risenhoover et al. [313] demonstrated that PA greatly reduced the digestion of cellulose because PA was toxic to rumen microbes. These experiments are evidence that PA is toxic to both caecalids (snowshoe hare) and the microbes that ruminants such as the wapiti rely on for their digestion of woody browse. Thus, these experiments suggest that PA is a broad-spectrum antifeedent toxin in the case of the browsing mammals of the North American boreal forest.

The lack of the resin glands that produce PA [349] on the tips of juvenile *B. papyrifera* twigs as compared to juvenile *B. neoalaskana* twigs (Figure 8.2) should reduce PA production by *B. papyrifera* twigs as compared to PA production by the tips of the twigs of juvenile *B. neoalaskana*, and this is the case. The current-annual-growth (CAG) twigs of

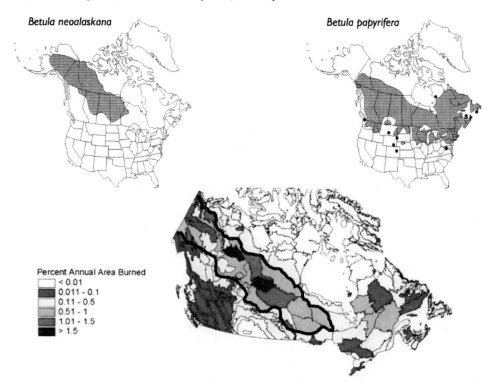

FIGURE 8.1: The distributions of the Alaska resin birch *Betula neoalaskana* and the white birch *B. papyrifera* and the percent area annually burned (PAAB) in Canada from 1959–1997 by large wildfires (fires > 200 ha [345]). In the PAAB map a black line outlines *B. neoalaskana*'s distribution, which encompasses the highest PAAB region of western Canada. The distribution maps of *B. neoalaskana* [270] and *B. papyrifera* [271] were provided courtesy of the Flora of North America Association. The PAAB map was provided courtesy of Mike Flannigan.

both the seedlings and the saplings of *B. papyrifera* grown in a birch garden in Finland from seed collected in Michigan contained no PA [180]. In contrast, both the seedlings and the saplings of *B. neoalaskana* grown at the same time in the same Finnish birch garden from seed collected at Fairbanks, Alaska, contained more than enough PA in their CAG twigs to deter browsing by snowshoe hares [180]: seedling PA concentration range 3.39% – 6.60%; sapling PA concentration $12 \pm 2\%$, mean ± 1 SEM; a PA concentration of 2% dry mass deterred feeding by Alaskan snowshoe hares [307]. In an extensive survey of PA production by wild saplings and its relationship to PAAB Bryant et al. [39] found that the average dry mass concentration of PA in the CAG twigs of 304 *B. neoalaskana* saplings (23% $\pm$ 0.7%, mean ± 1 SEM, range 7% dry mass – 88% dry mass) collected over the entire range of *B. neoalaskana* was significantly (t-test $p < 0.0001$) greater than the average dry mass concentration of PA in the CAG twigs of 90 *B. papyrifera* saplings (0.06% $\pm$ 0.6%, mean ± 1 SEM, range 0.0% dry mass – 1.9% dry mass) collected in the Great Lakes region.

As pointed out above, feeding trials using snowshoe hares have consistently demonstrated that in winter these herbivores eat significantly more biomass of the juvenile *B. papyrifera* than the twigs of *B. neoalaskana* [44]. Since in winter the browse preferences of snowshoe hares is similar to the browse preferences of moose (*Alces alces*) and most other North American boreal browsing mammals [42], these feeding trials using snowshoe hares

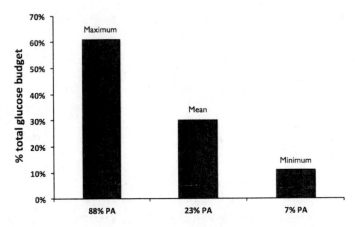

FIGURE 8.2: The cost of antibrowsing defense that the current-year-growth (CAG) twigs of *Betula neoalaskana* pay for producing papyriferic acid (PA) as estimated by the percentage of CAG twigs' total glucose budget consumed by PA production computed by the method of Gershenzon [123]. From left to right, the x-axis labels are the maximum PA dry mass concentration (PADMC) measured in the CAG twigs of a single sapling (88%), the mean PADMC of the 304 *B. neoalaskana* saplings assayed for PA (23%), and the minimum PADMC measured in the CAG twigs of a single sapling (7%).

strongly indicate that most browsing mammals of the North American boreal forest will preferentially browse *B. papyrifera* recruitment more severely that *B. neoalaskana* recruitment, and furthermore, that the cause of this differential selective browsing is a direct result of the lower production of PA by *B. papyrifera* as compared to *B. neoalaskana* [39, 180]. Thus, an obvious benefit of increased PA production by *B. neoalaskana* as compared to *B. papyrifera* in high PAAB regions that are often characterized by a high biomass of browsing mammals [130, 266] is the reduced browsing of *B. neoalaskana* as compared to *B. papyrifera* that we have predicted could facilitate an invasion of *B. neoalaskana* into the range of *B. papyrifera* in high PAAB regions.

However, the production of terpenoids such as PA also incurs a cost that could affect the ability of *B. neoalaskana* to invade into the range of *B. papyrifera*. Due to their extensive reduction, terpenoids like PA that defend the twigs of the juveniles of winter-dormant *B. neoalaskana* against browsing require more glucose to produce than the condensed tannins [123] and are the dominant secondary metabolite of the twigs of the juveniles winter-dormant *B. papyrifera* [180]. Moreover, the enzyme costs of making terpenoids are also comparatively high [124], and because of their toxicity [247] terpenoids are usually sequestered in complex, multicellular structures like *B. neoalaskana* resin glands, so storage costs for these substances are also likely to be substantial [124]. Thus, the increased production of PA by *B. neoalaskana* as compared to the minimal or no production of PA by *B. papyrifera* should reduce the growth rate of juvenile *B. neoalaskana* as compared to juvenile *B. papyrifera*, and this appears to be the case in birch provenance gardens located in British Columbia (C. Hawkins per. comm.).

The "take-home message" of the above paragraphs is twofold. First, increased production of papyriferic acid can be beneficial where browsing of forest recruitment is significant because by reducing browsing papyriferic acid reduces the decrease in growth and competitive ability that is caused by browsing. But where browsing is a less significant control over the growth and competitive ability of forest recruitment the cost of increased production of papyriferic acid, or for that matter any costly secondary metabolite [123, 124], may be sufficiently high that it reduces growth and competitive ability to the extent that defense production is selected against.

In short, theoretical analysis of the effect that toxin-dependent selective browsing has on plant biogeography, in this case of the biogeography of the two tree birches *B. neoalaskana* and *B. papyrifera*, is a cost-benefit problem. And to address this problem requires knowledge of the ecological stage on which the plant-herbivore play is acted. In the case of the North American boreal forest this requires an understanding of the role that wildfire plays in browsing mammal population dynamics.

8.4   Fire's relationship to the distributions of *B. neoalaskana* and *B. papyrifera*

Fire has been the dominant disturbance in the North American boreal forest since this forest's inception at the end of the last Ice Age [345]. In the boreal forest of North America fire is essential for the very existence of major early successional trees such *B. neoalaskana* and *B. papyrifera*, and is responsible for shaping landscape diversity and the physiognomy of this biome [144]. The fire mosaic of the North American boreal forest is at least 6,000 years old [103], and the region of highest percentage of area annually burned (PAAB) in this mosaic currently occurs in northwest Canada ([345]; see the lower map in Figure 8.1).

The lower map in Figure 8.1 shows the geographic variation in PAAB caused by large wildfires (> 200 hectares) that occurred in Canada from 1959–1997 [345]. The distributions of *B. neoalaskana* and *B. papyrifera* in western Canada are shown in the upper maps of Figure 8.1. In concert, these three maps show that *B. neoalaskana*'s distribution is more closely associated with regions of high PAAB than *B. papyrifera*'s distribution.

We tested this hypothesis by computing the average value of PAAB in the locations of Janet Dugle's ([90]) 76 *B. neoalaskana* collections and comparing it to the average value of PAAB in the locations of her 54 *B. papyrifera* collections (Figure 8.3). As predicted, we found that within the range of *B. neoalaskana* the average value of PAAB ($0.348\% \pm 0.051\%$, mean ± 1 SEM) was significantly (t-test, $p = 0.004$) higher than the average value of PAAB within the range of *B. papyrifera* ($0.143\% \pm 0.054\%$, mean ± 1 SEM).

Moreover, the greater production of PA by *B. neoalaskana* provenances from regions of high PAAB as compared to the lesser production of PA by *B. neoalaskana* provenances from regions of lower PAAB [39] should increase the glucose cost of defense (Figure 8.2) thereby reducing the growth rate of the juveniles of *B. neoalaskana* provenances from higher PAAB regions below the growth rate of the juveniles of *B. neoalaskana* provenances from lower PAAB regions. This prediction has recently been verified in a common garden experiment [344].

These considerations lead to the prediction that the production of PA in the recruitment of *B. neoalaskana* and *B. papyrifera* is most likely to be of highest selective value in the high fire regions of northwest Canada and of the least selective value in the lower fire regions found to the south and to the east (see lower map in Figure 8.1). The TDFRM in Chapter 4 described competition between two plant species with tradeoffs, such that the more slowly growing species was more toxic to herbivores. The slower growing species would be predicted by the model to have higher fitness in competing with the faster growing species only under conditions of high herbivory. This is also illustrated in Figure 8.4, which shows two plots of the simulation results of the TDFRM (7.1). The two plant species have different growth rates and toxicity levels, with plant species 1 having a higher growth rate ($r_1 > r_2$) and lower

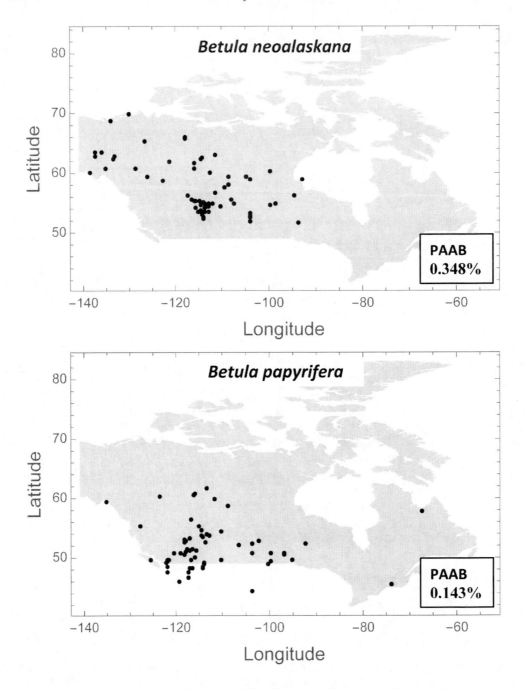

FIGURE 8.3: The locations of Dugle's [90] collections of *B. neoalaskana* and *B. papyrifera* and the average percent annual area burned (PAAB) by large wildfires (fires > 200 ha [345]). The average values of PAAB were computed by the method of Bryant et al. [39] using data in the Canada large Fire Database [345] and data in the United States Bureau of Land Management's Alaska Fire Service Data Base.

toxicity level $(G_1 > G_2)$. It shows that plant species 1 (species 2) dominates when herbivore density is low (high). The parameter values are: $r_1 = 0.0025, r_2 = 0.0015, G_1 = 98, G_2 =$

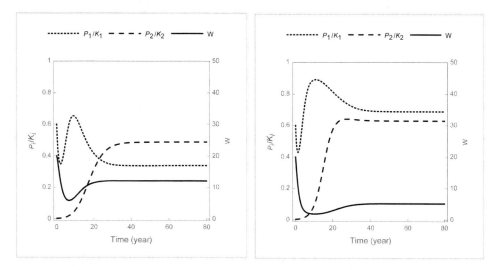

FIGURE 8.4: Simulations of the 3-D TDFRM (7.1) under high herbivore (left) and low herbivory (right) with plant species 1 having a higher growth rate and a lower plant toxicity level.

$5, \sigma_1 = 0.7, \sigma_2 = 0.3, c_{12} = 0.3, c_{21} = 0.1, h_1 = 0.01, h_2 = 0.05, e_1 = e_2 = 0.0002, B_1 = B_2 = 0.0003, B_w = 1, e_w = 0.00006, r_w = 0.001, \mu_w = 1/(10 \times 365), m_w = 0.0014, \mu_h = 1/(9 \times 365)$. The two herbivore densities in the two plots correspond to herbivore hunting rates $m_h = 0.0006$ (left) and 0.0014 (right).

Chapter 9

Example of *Mathematica* Notebooks

This chapter includes several examples of numerical simulations of model equations using the computation software *Mathematica*. The command "Manipulate" can be used to create a notebook that provides interactive visualizations, which allows us to view more easily the outcomes of the model as the model parameters change their values.

Examples of *Mathematica* notebooks are included at the end of this chapter in Section 9.4. These codes have been written based on version 10.0 of the software. Some of the commands may need to be modified if a newer version is used.

9.1   Simulations of a simple plant-herbivore model

The *Mathematica* notebook I titled "Simple model of plant-herbivore interactions" is a computer code that generates the figures shown in Figure 9.1. This simple model is described by the equations:

$$
\begin{aligned}
\frac{dP}{dt} &= rP\left(1 - \frac{P}{K}\right) - \frac{ePH}{1 + ehP} \\
\frac{dH}{dt} &= \frac{eBNH}{1 + ehN} - (\mu_h + m_h)H.
\end{aligned}
\tag{9.1}
$$

The variables P and H denoted the population sizes of plant and herbivore, respectively. The parameters are r, intrinsic growth rate of plant; K, plant carrying capacity; e, encounter rate; h, handling time; μ, natural mortality of herbivore; and d_h, herbivore hunting rate.

The panel on the left shows the ranges of several parameters whose values can be varied by sliding the bars within the range for each parameter (e.g., for r, K, h, B, m_p, etc.). The plots on the right show the updated figure when parameter values in the panel are changed. For example, Figures 9.1 (a–c) illustrate the outcomes for three values of the parameter m_h, which represent low ($m_h = 0.0006$), medium ($m_h = 0.001$), and high ($m_h = 0.003$) rates of herbivore hunting, respectively. Both the time plot (left) and the phase portrait (right) are presented. We observe that, for lower hunting rates, the system has a stable focus and solutions convergent to an interior equilibrium ((a) and (b)) with the plant density increasing with m_h, whereas herbivore extinction is expected when the hunting rate passes through a critical level, in which case the system converges to the boundary equilibrium $(P, H) = (K, 0)$ (see (c)).

(a)

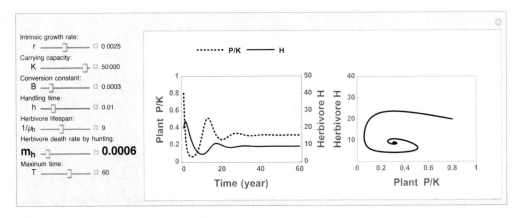

(b)

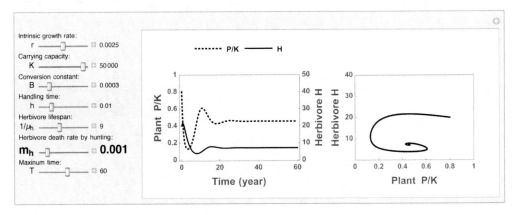

(c)

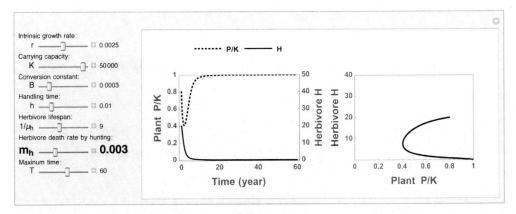

FIGURE 9.1: Simulation results of system (9.1) generated by **Notebook I**. It illustrates changes of the system from coexistence of plant and herbivore ((a) and (b)) to herbivore extinction (c) as the herbivore hunting rate m_h increases from 0.0006 to 0.001 and to 0.003. The values of other parameters are listed in Notebook I.

9.2   Sample codes for simulations of TDFRM

Notebook II provides a code for a 4-dimensional TDFRM with two plant species (P_1 and P_2), one herbivore population (H), and one wolf population (W). It is similar to the model (7.1) and has the following equations:

$$\frac{dP_i}{dt} = r_i P_i \left(1 - \frac{P_i + c_{ij}P_j}{K_i}\right) - C_i(P_1, P_2)H, \quad i,j = 1, 2, \ i \neq j,$$

$$\frac{dH}{dt} = \sum_{i=1}^{2} B_i C_i(P_1, P_2)H - e_w HW - (\mu_h + m_h)H, \tag{9.2}$$

$$\frac{dW}{dt} = B_w e_w HW - (d_w + m_w)W.$$

All parameters and the response function $C(P_1, P_2)$ have the same meanings as in model (7.1). Examples for the outcomes of the system (9.2) are illustrated in Figure 9.2.

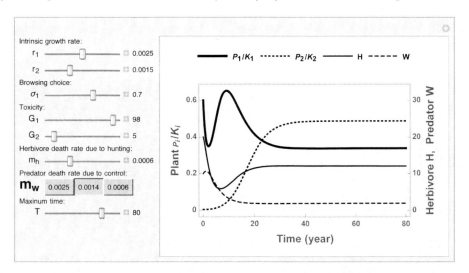

FIGURE 9.2: Simulation results of the model (9.2) generated by **Notebook II**. It shows the population densities in the case when the hunting rate of predator is $m_w = 0.0014$, which represents a predator control with 40% of decrease per year.

One of the parameters on the left panel that is varied in Figure 9.2 is the wolf hunting rate m_w. The plot on the right shows the updated solution behavior corresponding to the varied parameter values. For example, in Figure 9.2 the value for the parameter m_W is 0.0014, which represents that the predator (wolf) population is decreased by 40% per year through wolf hunting.

To explore other policy scenarios, different percentages of hunting can be considered by varying the death rate of predator due to hunting (m_w). For example, Figure 9.2 is for the case when all parameters are fixed except m_w. The two different values of m_w are listed in the panel, one is $m_w = 0.0025$ representing an increased predator control from 40% per year as shown in Figure 9.2 to 60% (see Figure 9.3a), and the other one is $m_w = 0.0006$ representing a decreased predator control from 40% to 20% (see Figure 9.3b). It shows that for higher level of wolf control, the herbivore population is relatively high leading to a relatively higher density of more toxic plant species (P_1), whereas a lower level of wolf control can reduce the herbivore population leading to a much higher density of less toxic plant species (P_2).

(a)

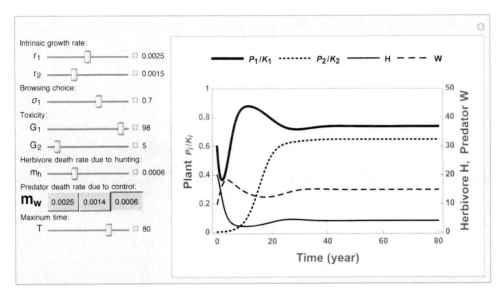

(b)

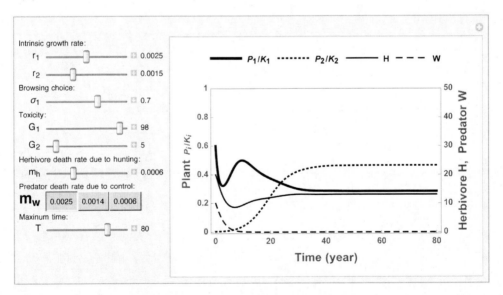

FIGURE 9.3: Similar to Figure 9.2 (generated by Notebook II) but for different values of predator control m_w. It shows simulation results of the model (9.2) when (a) $m_w = 0.0006$ (i.e., the predator control is decreased to 20% per year) or (b) $m_w = 0.0025$ or the predator control is increased to 60% per year.

Another parameter representing policy decisions is the death rate of herbivore (e.g., moose) due to hunting, m_h. Figure 9.4 shows simulation results using **Notebook III** for two different values of m_h. The wolf control rate m_w is fixed at 0.0014 and all other parameters have the same values as in Figure 9.2. Plot (a) is for the case of lower moose hunting rate $m_h = 0.0001$, whereas plot (b) is for a higher moose hunting rate $m_h = 0.0014$. It shows the dominance of more toxic plant species (plant 2) when hunting rate is lower and dominance of less toxic plant species (plant 1) when the hunting rate is higher.

(a)

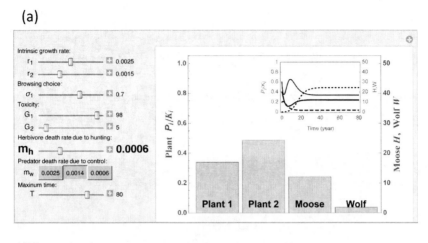

(b)

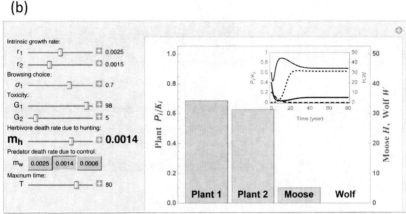

FIGURE 9.4: Similar to Figure 9.3 but with varying rates of herbivore hunting m_h, with bar chart presenting steady-state values. This is generated by **Notebook III**. Plot (a) is for low level of hunting $m_h = 0.0001$ and Plot (b) is for higher level of hunting $m_h = 0.0014$. The bar chart shows the densities of plant species 1 and 2, herbivore (moose), and predator (wolf). The insert shows time plots similar to Figure 9.3 (the legend is not shown).

9.3 Plots of stochastic simulations for spatial distributions of vegetation

Notebook IV provides a code for importing data generated by ALFRESCO simulations for spatial vegetation distributions and plotting the outcomes. Figure 9.5 illustrates the simulation outcomes. It compares three scenarios of wolf control described by $m_w = 0.0006$ (low) $m_w = 0.0014$ (medium), and $m_w = 0.0025$ (high). The simulation consists of 100 replicates, all of which can be viewed for vegetation changes from time $t = 0$ to $t = 92$ (see Notebook IV for the panels showing time, replicates, and scenarios). In the figures, spruce (green, or lighter color) and deciduous (black, or darker color) represent more toxic and less toxic plant species, respectively. Figure 9.5 illustrates the vegetation distributions at $t = 92$ under low (see(a)) and high (see (b)) predator control. It shows that a lower (higher) predator control leads to a higher (lower) percentage of deciduous.

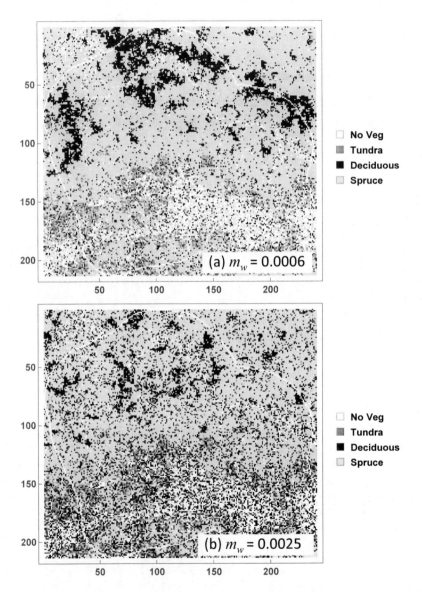

FIGURE 9.5: Results from ALFRESCO for spatially explicit stochastic simulations of vegetation dynamics in Tanana Flats from year 2007 to 2099. The figures are generated by **Notebook IV**. It compares the outcomes under low (A, $m_w = 0.0006$) and high (B, $m_w = 0.0025$) predator control scenarios.

9.4   *Mathematica* notebooks

I. Simple model of plant-herbivore interactions (used for Figure 9.1)

```
Manipulate[Module[{P, H, t, sol, eqa, el, μh, h, P0, H0, figPHtime, figPHphase, pmax,
   f, MaxTime, year, Pbd},

  h = 0.01; el = 0.0002; μh = 1/(s year); P0 = .8 k; H0 = 20; MaxTime = T year;
  year = 365; Pbd = 1; f = el P[t] / (1 + h el P[t]);

  eqa = {P'[t] == r P[t] (1 - P[t] / k) - H[t] f, H'[t] == B H[t] f - (μh + mh) H[t],
       P[0] == P0, H[0] == H0};
  sol = NDSolve[eqa, {P, H}, {t, 0, MaxTime}];

  figPHtime = Plot[Evaluate[{P[t] / k, H[t] / 50} /. sol], {t, 0, MaxTime},
     PlotRange → {-0.01, Pbd}, PlotStyle → {{Red, Dotted, Thick}, {Blue}},
     Frame → True, FrameTicks → {{{0, .2, .4, .6, .8, 1}, {0, {.2, "10"}, {.4, "20"},
        {.6, "30"}, {.8, "40"}, {1, "50"}}}, {{{0, "0"}, {20 year, "20"}, {40 year, "40"},
        {60 year, "60"}, {80 year, "80"}, {100 year, "100"}, {150 year, "150"},
        {200 year, "200"}, {300 year, "300"}}, None}}, AspectRatio → .7, ImageSize → 240,
     FrameLabel → {"Time (year)", "Plant  P/K", "", "Herbivore H"},
     PlotLegends → Placed[{"P/K", "H"}, Above], LabelStyle → {Bold, 10}];

  figPHphase = ParametricPlot[Evaluate[{P[t] / k, H[t]} /. sol], {t, 0, MaxTime},
     PlotRange → {{0, 1}, {0, 2 H0}}, PlotStyle → {Purple, Thick}, Frame → True,
     FrameTicks → {{{10, 20, 30, 40, 50}, None}, {{0, .2, .4, .6, .8, 1}, None}},
     FrameLabel → {"Plant  P/K", "Herbivore H", "", ""}, AspectRatio → .7,
     ImageSize → 220, LabelStyle → {Bold, 10}];

  Row[{figPHtime, Space, figPHphase}]
],

Style["Intrinsic growth rate:"],
{{r, 0.0025, Style["r", 11]}, 0, 0.005, 0.0001, Appearance → "Labeled", ImageSize → Tiny},
Style["Carrying capacity:"],
{{k, 50 000, Style["K", 11]}, 0, 50 000, 1000, Appearance → "Labeled", ImageSize → Tiny},
Style["Conversion constant:"],
{{B, 0.0003, Style["B", 11]}, 0, .002, 0.00001, Appearance → "Labeled", ImageSize → Tiny},
Style["Handling time:"],
{{h, 0.01, Style["h", 11]}, 0, .05, 0.0001, Appearance → "Labeled", ImageSize → Tiny},
Style["Herbivore lifespan:"],
{{s, 9, Style["1/μh", 11]}, 5, 15, 1, Appearance → "Labeled", ImageSize → Tiny},
Style["Herbivore death rate by hunting:"],
{{mh, .0006, Style["mh", Bold, 16]}, 0, 0.01, 0.0001, Appearance → "Labeled",
   LabelStyle → {Bold, 16}, ImageSize → Tiny},
Style["Maximum time:"],
{{T, 60, Style["T", 11]}, 1, 100, 1, Appearance → "Labeled", ImageSize → Tiny}
]
```

II. TDFRM with two-plant, herbivore, and predator (for Figures 9.2 and 9.3)

```
Manipulate[Module[{P1, P2, H, W, t, sol, eqa, e1, e2, σ2, ew, μh, μw, rw, K1, K2, c12, c21, h1,
   h2, B1, B2, Bw, Kw, P10, P20, H0, W0, fig, figPHWtime, pmax, f1, f2, MaxTime, year, Pbd, Hbd},

  e1 = e2 = 0.0002; σ2 = 1 - σ1; ew = 0.00006; rw = 0.001; c12 = 0.3; c21 = 0.1; μh = 1 / (9 year);
  μw = 1 / (10 year); h1 = 0.01; h2 = 0.05; B1 = B2 = 0.0003; Bw = 1; K1 = 50 000; K2 = 50 000;
  Kw = 40; P10 = .6 K1; P20 = 200; H0 = 20; W0 = 10; MaxTime = T year; year = 365; Pbd = 1; Hbd = 50;
  f1 = e1 σ1 P1[t] / (1 + h1 e1 σ1 P1[t] + h2 e2 σ2 P2[t]);
  f2 = e2 σ2 P2[t] / (1 + h1 e1 σ1 P1[t] + h2 e2 σ2 P2[t]);

  eqa = {P1'[t] ⩵ r1 ∗ P1[t] (1 - (P1[t] + c21 ∗ P2[t]) / K1) - H[t] f1 (1 - f1 / (4 G1)),
      P2'[t] ⩵ r2 ∗ P2[t] (1 - (P2[t] + c12 ∗ P1[t]) / K2) - H[t] f2 (1 - f2 / (4 G2)),
    H'[t] ⩵ B1 H[t] f1 (1 - f1 / (4 G1)) + B2 H[t] f2 (1 - f2 / (4 G2)) - ew H[t] W[t] - (μh + mh) H[t],
    W'[t] ⩵ Bw ew H[t] W[t] + rw (1 - W[t] / Kw) W[t] - (μw + mw) W[t],
    P1[0] ⩵ P10, P2[0] ⩵ P20, H[0] ⩵ H0, W[0] ⩵ W0};

  sol = NDSolve[eqa, {P1, P2, H, W}, {t, 0, MaxTime}];

  figPHWtime = Plot[Evaluate[{P1[t] / K1, P2[t] / K2, H[t] / 50, W[t] / 50} /. sol], {t, 0, MaxTime},
    PlotRange → {-0.01, Pbd}, PlotStyle → {{Purple, Thickness[.01]}, {Red, Thick, Dotted},
      {Blue}, {Brown, Dashing[0.02]}}, Frame → True, FrameTicks → {{{0, .2, .4, .6, .8, 1},
      {0, {.2, "10"}, {.4, "20"}, {.6, "30"}, {.8, "40"}, {1, "50"}}},
      {{{0, "0"}, {20 year, "20"}, {40 year, "40"}, {60 year, "60"}, {80 year, "80"}}, None}},
    PlotLegends → Placed[{"P₁/K₁", "P₂/K₂", "H", "W"}, Above], ImageSize → 350,
    FrameLabel → {"Time (year)", "Plant Pᵢ/Kᵢ", "", "Herbivore H,  Predator W"},
    LabelStyle → {Bold, 10}
   ];

  Row[{figPHWtime}]
 ],

Style["Intrinsic growth rate:"],
{{r1, 0.0025, Style["r₁", 11]}, 0, 0.005, 0.001, Appearance → "Labeled", ImageSize → Small},
{{r2, 0.0015, Style["r₂", 11]}, 0, 0.005, 0.001, Appearance → "Labeled", ImageSize → Small},
Style["Browsing choice:"],
{{σ1, 0.7, Style["σ₁", 11]}, 0.1, 1, 0.1, Appearance → "Labeled", ImageSize → Small},
Style["Toxicity:"],
{{G1, 98, Style["G₁", 11]}, 0, 100, 1, Appearance → "Labeled", ImageSize → Small},
{{G2, 5, Style["G₂", 11]}, 0, 100, 1, Appearance → "Labeled", ImageSize → Small},
Style["Herbivore death rate due to hunting:"],
{{mh, 0.0006, Style["m₍ₕ₎", 11]}, 0, 0.002, 0.0001, Appearance → "Labeled", ImageSize → Small},
Style["Predator death rate due to control:"],
{{mw, 0.0014, Style["m₍w₎", Bold, 18]}, {0.0025, 0.0014, 0.0006}},
Style["Maximum time:"],
{{T, 80, Style["T", 11]}, 1, 100, 1, Appearance → "Labeled", ImageSize → Small}
 ]
```

III. TDFRM with two-plant, herbivore, and predator (for Figure 9.4)

```
Manipulate[Module[{P1, P2, H, W, t, sol, eqa, e1, e2, σ2, ew, μh, μw, rw, K1, K2, c12, c21, h1,
    h2, B1, B2, Bw, Kw, P10, P20, H0, W0, pmax, f1, f2, MaxTime, Hbd, SteadyState, fig1, fig2},

  e1 = e2 = 0.0002; σ2 = 1 - σ1; ew = 0.00006; rw = 0.001; c12 = 0.3; c21 = 0.1; μh = 1 / (9 * 365);
  μw = 1 / (10 * 365); h1 = 0.01; h2 = 0.05; B1 = 0.0003; B2 = 0.0003; Bw = 1; K1 = 50 000;
  K2 = 50 000; Kw = 40; P10 = .6 K1; P20 = 200; H0 = 20; W0 = 10; MaxTime = 365 T; Hbd = 50;
  f1 = e1 σ1 P1[t] / (1 + h1 e1 σ1 P1[t] + h2 e2 σ2 P2[t]);
  f2 = e2 σ2 P2[t] / (1 + h1 e1 σ1 P1[t] + h2 e2 σ2 P2[t]);

  eqa = {P1'[t] == r1 * P1[t] (1 - (P1[t] + c21 * P2[t]) / K1) - H[t] f1 (1 - f1 / (4 G1)),
    P2'[t] == r2 * P2[t] (1 - (P2[t] + c12 * P1[t]) / K2) - H[t] f2 (1 - f2 / (4 G2)),
    H'[t] == B1 H[t] f1 (1 - f1 / (4 G1)) + B2 H[t] f2 (1 - f2 / (4 G2)) - ew H[t] W[t] - (μh + mh) H[t],
    W'[t] == Bw ew H[t] W[t] + rw (1 - W[t] / Kw) W[t] - (μw + mw) W[t],
    P1[0] == P10, P2[0] == P20, H[0] == H0, W[0] == W0};
  sol = NDSolve[eqa, {P1, P2, H, W}, {t, 0, MaxTime}];
  SteadyState = Evaluate[{P1[365 T] / K1, P2[365 T] / K2, H[365 T] / 50, W[365 T] / 50} /. sol];
  fig1 = Plot[Evaluate[{P1[t] / K1, P2[t] / K2, H[t] / 50, W[t] / 50} /. sol], {t, 0, MaxTime},
    PlotRange → {-0.01, 1}, PlotStyle → {{Black, Thickness[.01]}, {Black, Dotted},
      {Black}, {Black, Dashed}}, Frame → True, FrameTicks → {{{0, .2, .4, .6, .8, 1},
      {0, {.2, "10"}, {.4, "20"}, {.6, "30"}, {.8, "40"}, {1, "50"}}},
      {{0, {20 * 365, "20"}, {40 * 365, "40"}, {60 * 365, "60"}, {80 * 365, "80"}}, None}},
    ImageSize → 350, FrameLabel → {"Time (year)", "P_i/K_i", "", "H,W"}, LabelStyle → {8}];
  fig2 = BarChart[First[SteadyState], PlotRange → {0, 1}, ImageSize → {200, 400},
    ChartLabels → Placed[{"Plant 1", "Plant 2", "Moose", "Wolf"}, Bottom],
    AspectRatio → .8, Frame → True, ChartStyle → LightGray,
    FrameTicks → {{Automatic, {0, {.2, "10"}, {.4, "20"}, {.6, "30"}, {.8, "40"}, {1, "50"}}},
      {{{0, "0"}, {20 * 365, "20"}, {40 * 365, "40"}, {60 * 365, "60"}, {80 * 365, "80"}}, None}},
    FrameLabel → {"", "Plant   P_i/K_i", "", "Moose H, Wolf W"}, LabelStyle → {Bold, 10}];

  Graphics[{Inset[fig2, {0, .7}, Center, {4, 6}], Inset[fig1, {0.6, 1.4}, Center, {2, 2}]},
    PlotRange → 2]
  ],

Style["Intrinsic growth rate:"],
{{r1, 0.0025, Style["r_1", 11]}, 0, 0.005, 0.001, Appearance → "Labeled", ImageSize → Small},
{{r2, 0.0015, Style["r_2", 11]}, 0, 0.005, 0.001, Appearance → "Labeled", ImageSize → Small},
Style["Browsing choice:"],
{{σ1, 0.7, Style["σ_1", 11]}, 0.1, 1, 0.1, Appearance → "Labeled", ImageSize → Small},
Style["Toxicity:"],
{{G1, 98, Style["G_1", 11]}, 0, 100, 1, Appearance → "Labeled", ImageSize → Small},
{{G2, 5, Style["G_2", 11]}, 0, 100, 1, Appearance → "Labeled", ImageSize → Small},
Style["Herbivore death rate due to hunting:"],
{{mh, 0.0006, Style["m_h", Bold, 18]}, 0, 0.002, 0.0001, Appearance → "Labeled",
  LabelStyle → {Bold, 16}, ImageSize → Small},
Style["Predator death rate due to control:"],
{{mw, 0.0014, Style["m_w", 11]}, {0.0025, 0.0014, 0.0006}},
Style["Maxinum time:"],
{{T, 80, Style["T", 11]}, 1, 100, 1, Appearance → "Labeled", ImageSize → Small}
]
```

IV. Spatial maps of simulation results

```
Manipulate[Module[{data, inputDirectory, map, border, bordermap},
    inputDirectory = "/Users/fengz/Desktop/Output/data_mw";
  data = Import[StringJoin[inputDirectory, StringDrop[ToString[mw], 2],
      "/Maps/", "Veg_", IntegerString[rep], "_",
      IntegerString[year], ".txt"], "Table"][[7 ;; 219]];
  border = Import["/Users/fengz/Desktop/Output/Tananaflats_Border.txt",
    "Table"]; bordermap = MatrixPlot[border,
    ColorRules → {1 → {White, Thick}, -9999 → Transparent}];
  map = MatrixPlot[data, ColorRules → {0 → White, 1 → Orange, 2 → Green,
      3 → Green, 4 → RGBColor[0, 0.2, 0]}, PlotRange → {{0, 250}, {0, 250}},
    LabelStyle → {Bold, 11}, ImageSize → 350,
    Frame → True, FrameTicks → {Automatic, Automatic, None, None},
    PlotLegends → {"No Veg", "Tundra", "Deciduous", "Spruce"}];

  Show[map, bordermap]],

{{mw, 0.0014, Style["mw=", Bold, 12]}, {0.0006, 0.0014, 0.0025}},
{{rep, 50, Style["Replicate", Bold, 12]}, 0, 99, 1, Appearance → "Labeled"},
{{year, 1, Style["Year", Bold, 12]}, 0, 92, 1, Appearance → "Labeled"}]
```

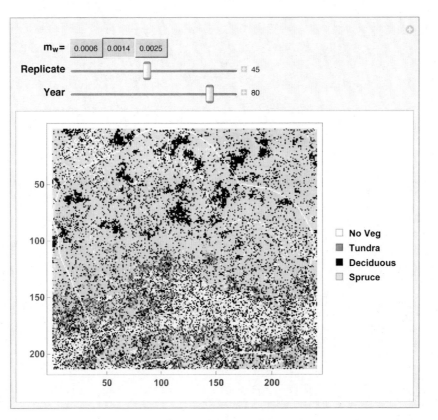

References

[1] K.C. Abbott and G. Dwyer. Food limitation and insect outbreaks: complex dynamics in plant-herbivore models. *Journal of Animal Ecology*, 76(5):1004–1014, 2007.

[2] P.A. Abrams. Functional responses of optimal foragers. *The American Naturalist*, 120(3):382–390, 1982.

[3] P.A. Abrams. Foraging time optimization and interactions in food webs. *The American Naturalist*, 124(1):80–96, 1984.

[4] P.A. Abrams. The effects of adaptive behavior on the type-2 functional response. *Ecology*, 71(3):877–885, 1990.

[5] P.A. Abrams. Life history and the relationship between food availability and foraging effort. *Ecology*, 72(4):1242–1252, 1991.

[6] P.A. Abrams. Adaptive foraging by predators as a cause of predator-prey cycles. *Evolutionary Ecology*, 6(1):56–72, 1992.

[7] P.A. Abrams. Effect of increased productivity on the abundances of trophic levels. *The American Naturalist*, 141(3):351–371, 1993.

[8] P.A. Abrams. Why predation rate should not be proportional to predator density. *Ecology*, 74(3):726–733, 1993.

[9] P.A. Abrams and C.J. Walters. Invulnerable prey and the paradox of enrichment. *Ecology*, 77(4):1125–1133, 1996.

[10] A.A. Agrawal. Phenotypic plasticity in the interactions and evolution of species. *Science*, 294(5541):321–326, 2001.

[11] A.A. Agrawal and R. Karban. *Why induced defenses may be favored over constitutive strategies in plants.* Princeton University Press, 1999.

[12] G.I. Ågren and O. Franklin. Root: shoot ratios, optimization and nitrogen productivity. *Annals of Botany*, 92(6):795–800, 2003.

[13] L.J.S. Allen, M.K. Hannigan, and M.J. Strauss. Mathematical analysis of a model for a plant-herbivore system. *Bulletin of Mathematical Biology*, 55(4):847–864, 1993.

[14] B.R. Anholt and E.E. Werner. Density-dependent consequences of induced behavior. In *The ecology and evolution of inducible defenses*, pages 218–230. Princeton University Press, 1999.

[15] R. Arditi, C. Lobry, and T. Sari. Is dispersal always beneficial to carrying capacity? New insights from the multi-patch logistic equation. *Theoretical Population Biology*, 106:45–59, 2015.

[16] T.S. Aronen. Interactions between agrobacterium tumefaciens and coniferous defence compounds α-pinene and trans-stilbene. *Forest Pathology*, 27(1):55–67, 1997.

[17] D.J. Augustine, L.E. Frelich, and P.A. Jordan. Evidence for two alternate stable states in an ungulate grazing system. *Ecological Applications*, 8(4):1260–1269, 1998.

[18] W.L. Baker. Effects of settlement and fire suppression on landscape structure. *Ecology*, 73(5):1879–1887, 1992.

[19] W.L. Baker. Spatially heterogeneous multi-scale response of landscapes to fire suppression. *Oikos*, 66(1):66–71, 1993.

[20] E.S. Bakker, R.C. Reiffers, H. Olff, and J.M. Gleichman. Experimental manipulation of predation risk and food quality: effect on grazing behaviour in a central-place foraging herbivore. *Oecologia*, 146(1):157–167, 2005.

[21] D. Baldocchi, F.M. Kelliher, A.T. Black, and P.l Jarvis. Climate and vegetation controls on boreal zone energy exchange. *Global Change Biology*, 6(S1):69–83, 2000.

[22] I.T. Baldwin and E.A. Schmelz. Immunological "memory" in the induced accumulation of nicotine in wild tobacco. *Ecology*, 77(1):236–246, 1996.

[23] W.B. Ballard, J.S. Whitman, and C.L. Gardner. Ecology of an exploited wolf population in south-central Alaska. *Wildlife Monographs*, (98):3–54, 1987.

[24] J.R. Beddington. Mutual interference between parasites or predators and its effect on searching efficiency. *The Journal of Animal Ecology*, 44(1):331–340, 1975.

[25] G.E. Belovsky. Diet optimization in a generalist herbivore: the moose. *Theoretical Population Biology*, 14(1):105–134, 1978.

[26] G.E. Belovsky and O.J. Schmitz. Plant defenses and optimal foraging by mammalian herbivores. *Journal of Mammalogy*, 75(4):816–832, 1994.

[27] U.A. Bergvall, P. Rautio, K. Kesti, J. Tuomi, and O. Leimar. Associational effects of plant defences in relation to within-and between-patch food choice by a mammalian herbivore: neighbour contrast susceptibility and defence. *Oecologia*, 147(2):253–260, 2006.

[28] A.A. Berryman. Towards a unified theory of plant defense. In *Mechanisms of woody plant defenses against insects*, pages 39–55. Springer, 1988.

[29] K. Boege and R.J. Marquis. Facing herbivory as you grow up: the ontogeny of resistance in plants. *Trends in Ecology & Evolution*, 20(8):441–448, 2005.

[30] T.A. Bookhout. *The snowshoe hare in upper Michigan: its biology and feeding coactions with white-tailed deer*. Number 38. Michigan Department of Conservation, 1965.

[31] R. Boonstra, C.J. Krebs, and N.C. Stenseth. Population cycles in small mammals: the problem of explaining the low phase. *Ecology*, 79(5):1479–1488, 1998.

[32] D.B. Botkin and L.G. Simpson. Biomass of the North American boreal forest: a step toward accurate global measures. *Biogeochemistry*, 9(2):161–174, 1990.

[33] R. Boyle, S. McLean, W.J. Foley, and N.W. Davies. Comparative metabolism of dietary terpene, p-cymene, in generalist and specialist folivorous marsupials. *Journal of Chemical Ecology*, 25(9):2109–2126, 1999.

[34] R.R. Boyle, S. McLean, S. Brandon, and N. Wiggins. Rapid absorption of dietary 1, 8-cineole results in critical blood concentration of cineole and immediate cessation of eating in the common brushtail possum (*Trichosurus vulpecula*). *Journal of Chemical Ecology*, 31(12):2775–2790, 2005.

[35] J.S. Brown and B.P. Kotler. Hazardous duty pay and the foraging cost of predation. *Ecology Letters*, 7(10):999–1014, 2004.

[36] J.P. Bryant. The regulation of snowshoe hare feeding behavior during winter by plant antiherbivore chemistry. *Proceedings of the International Lagomorph Conferences*, 1981.

[37] J.P. Bryant and F.S. Chapin III. Browsing-woody plant interactions during boreal forest plant succession. In *Forest ecosystems in the Alaskan taiga*, pages 213–225. Springer, 1986.

[38] J.P. Bryant, F.S. Chapin III, and D.R. Klein. Carbon/nutrient balance of boreal plants in relation to vertebrate herbivory. *Oikos*, 40(3):357–368, 1983.

[39] J.P. Bryant, T.P. Clausen, R.K. Swihart, S.M. Landhäusser, M.T. Stevens, C.D.B. Hawkins, S. Carrière, A.P. Kirilenko, A.M. Veitch, R.A. Popko, et al. Fire drives transcontinental variation in tree birch defense against browsing by snowshoe hares. *The American Naturalist*, 174(1):13–23, 2009.

[40] J.P. Bryant, K. Danell, F.D. Provenza, P.B. Reichardt, T.A. Clausen, and R.A. Werner. Effects of mammal browsing on the chemistry of deciduous woody plants. In *Phytochemical induction by herbivores*, pages 135–154. Wiley, New York, 1991.

[41] J.P. Bryant et al. Woody plant-mammal interactions. In *Herbivores: their interaction with secondary plant metabolites*. Academic Press, 1991.

[42] J.P. Bryant and P.J. Kuropat. Selection of winter forage by subarctic browsing vertebrates: the role of plant chemistry. *Annual Review of Ecology and Systematics*, 11(1):261–285, 1980.

[43] J.P. Bryant, F.D. Provenza, J. Pastor, P.B. Reichardt, T.P. Clausen, and J.T. du Toit. Interactions between woody plants and browsing mammals mediated by secondary metabolites. *Annual Review of Ecology and Systematics*, 22(1):431–446, 1991.

[44] J.P. Bryant, R.K. Swihart, P.B. Reichardt, and L. Newton. Biogeography of woody plant chemical defense against snowshoe hare browsing: comparison of Alaska and eastern North America. *Oikos*, 70(3):385–395, 1994.

[45] J.P. Bryant, G.D. Wieland, T. Clausen, and P. Kuropat. Interactions of snowshoe hare and feltleaf willow in Alaska. *Ecology*, 66(5):1564–1573, 1985.

[46] J.P. Bryant, G.D. Wieland, P.B. Reichardt, V.E. Lewis, and M.C. McCarthy. Pinosylvin methyl ether deters snowshoe hare feeding on green alder. *Science*, 222(4627):1023–1025, 1983.

[47] S.L. Buchmann and G.P. Nabhan. *The forgotten pollinators*. Island Press, 2012.

[48] T.N. Buckley and D.W. Roberts. DESPOT, a process-based tree growth model that allocates carbon to maximize carbon gain. *Tree Physiology*, 26(2):129–144, 2006.

[49] T. Burnett. Effects of temperature and host density on the rate of increase of an insect parasite. *The American Naturalist*, 85(825):337–352, 1951.

[50] L.G. Butler and K. Kielland. Acceleration of vegetation turnover and element cycling by mammalian herbivory in riparian ecosystems. *Journal of Ecology*, 96(1):136–144, 2008.

[51] L.G. Butler, K. Kielland, S.T. Rupp, and T.A. Hanley. Interactive controls of herbivory and fluvial dynamics on landscape vegetation patterns on the Tanana River floodplain, interior Alaska. *Journal of Biogeography*, 34(9):1622–1631, 2007.

[52] S.R. Cantrell, C. Cosner, D.L. DeAngelis, and V. Padron. The ideal free distribution as an evolutionarily stable strategy. *Journal of Biological Dynamics*, 1(3):249–271, 2007.

[53] C. Castillo-Chavez, Z. Feng, and W. Huang. Global dynamics of a plant-herbivore model with toxin-determined functional response. *SIAM Journal on Applied Mathematics*, 72(4):1002–1020, 2012.

[54] G. Caughley. Wildlife management and the dynamics of ungulate populations. *Applied Biology*, 1:183–246, 1976.

[55] G. Caughley and J.H. Lawton. Plant-herbivore systems. In *Theoretical ecology*, pages 132–166. Blackwell, Oxford, 1981.

[56] C.C. Celimene, J.A. Micales, L. Ferge, and R.A. Young. Efficacy of pinosylvins against white-rot and brown-rot fungi. *Holzforschung*, 53(5):491–497, 1999.

[57] F.S. Chapin III. The mineral nutrition of wild plants. *Annual Review of Ecology and Systematics*, 11:233–260, 1980.

[58] F.S. Chapin III, A.J. Bloom, C.B. Field, and R.H. Waring. Plant responses to multiple environmental factors. *Bioscience*, 37(1):49–57, 1987.

[59] F.S. Chapin III, T. Hollingsworth, D.F. Murray, L.A. Viereck, and M.D. Walker. *Floristic diversity and vegetation distribution in the Alaskan boreal forest*. Oxford University Press, New York, 2006.

[60] F.S. Chapin III, S. Rupp, A.M. Starfield, L. DeWilde, E.S. Zavaleta, N. Fresco, J. Henkelman, and D.A. McGuire. Planning for resilience: modeling change in human-fire interactions in the Alaskan boreal forest. *Frontiers in Ecology and the Environment*, 1(5):255–261, 2003.

[61] F.S. Chapin III, P.R. Tryon, and K.V. Cleve. Influence of phosphorus on growth and biomass distribution of Alaskan taiga tree seedlings. *Canadian Journal of Forest Research*, 13(6):1092–1098, 1983.

[62] F.S. Chapin III, L.A. Viereck, P. Adams, K. Van Cleve, C.L. Fastie, R.A. Ott, D. Mann, and J.F. Johnstone. Successional processes in the Alaskan boreal forest. In *Alaska's changing boreal forest*, pages 100–120. Oxford University Press New York, 2006.

[63] E.L. Charnov. Optimal foraging, the marginal value theorem. *Theoretical Population Biology*, 9(2):129–136, 1976.

[64] T.P. Clausen, J.P. Bryant, and P.B. Reichardt. Defense of winter-dormant green alder against snowshoe hares. *Journal of Chemical Ecology*, 12:2117–2131, 1986.

[65] K. Clay. Symbiosis and the regulation of communities. *American Zoologist*, 41(4):810–824, 2001.

[66] K. Clay and J. Holah. Fungal endophyte symbiosis and plant diversity in successional fields. *Science*, 285(5434):1742–1744, 1999.

[67] K. Clay, J. Holah, and J.A. Rudgers. Herbivores cause a rapid increase in hereditary symbiosis and alter plant community composition. *Proceedings of the National Academy of Sciences of the United States of America*, 102(35):12465–12470, 2005.

[68] K. Cleve, L.A. Viereck, and G.M. Marion. Introduction and overview of a study dealing with the role of salt-affected soils in primary succession on the Tanana River floodplain, interior Alaska. *Canadian Journal of Forest Research*, 23(5):879–888, 1993.

[69] Y. Cohen, J. Pastor, and R. Moen. Bite, chew, and swallow. *Ecological Modelling*, 116(1):1–14, 1999.

[70] P.D. Coley, J.P. Bryant, and F.S. Chapin III. Resource availability and plant antiherbivore defense. *Science*, 230(4728):895–899, 1985.

[71] H.N. Comins and M.P. Hassell. Predation in multi-prey communities. *Journal of Theoretical Biology*, 62(1):93–114, 1976.

[72] H.N. Comins and R.E. McMurtrie. Long-term response of nutrient-limited forests to $CO''2$ enrichment; equilibrium behavior of plant-soil models. *Ecological Applications*, 3(4):666–681, 1993.

[73] J.M. Craine. *Resource strategies of wild plants*. Princeton University Press, 2009.

[74] M.J. Crawley. Life history and environment. In *Plant ecology*, pages 73–131. Wiley Online Library, 2009.

[75] M.J. Crawley et al. *Herbivory. The dynamics of animal-plant interactions*. Blackwell Scientific Publications, 1983.

[76] R. Cressman and V. Křivan. The ideal free distribution as an evolutionarily stable state in density-dependent population games. *Oikos*, 119(8):1231–1242, 2010.

[77] S.G. Cumming. Forest type and wildfire in the Alberta boreal mixedwood: what do fires burn? *Ecological Applications*, 11(1):97–110, 2001.

[78] R.C. Dart. *Medical toxicology*. Lippincott Williams & Wilkins, 2004.

[79] H.J. De Knegt, G.M. Hengeveld, F. Van Langevelde, W.F. De Boer, and K.P. Kirkman. Patch density determines movement patterns and foraging efficiency of large herbivores. *Behavioral Ecology*, 18(6):1065–1072, 2007.

[80] C. De Mazancourt, M.l Loreau, and L. Abbadie. Grazing optimization and nutrient cycling: when do herbivores enhance plant production? *Ecology*, 79(7):2242–2252, 1998.

[81] D.L. DeAngelis, J.P. Bryant, R. Liu, S.A. Gourley, C.J. Krebs, and P.B. Reichardt. A plant toxin mediated mechanism for the lag in snowshoe hare population recovery following cyclic declines. *Oikos*, 124(6):796–805, 2015.

[82] D.L. DeAngelis, R.A. Goldstein, and R.V. O'neill. A model for tropic interaction. *Ecology*, 56(4):881–892, 1975.

[83] D.M. Dearing, W.J. Foley, and S. McLean. The influence of plant secondary metabolites on the nutritional ecology of herbivorous terrestrial vertebrates. *Annual Review of Ecology, Evolution, and Systematics*, pages 169–189, 2005.

[84] M.D. Dearing. Effects of Acomastylis rossii tannins on a mammalian herbivore, the North American pika, ochotona princeps. *Oecologia*, 109(1):122–131, 1996.

[85] R.C. Dewar. A root-shoot partitioning model based on carbon-nitrogen-water interactions and Munch phloem flow. *Functional Ecology*, 7(3):356–368, 1993.

[86] J.R. Donaldson, M.T. Stevens, H.R. Barnhill, and R.L. Lindroth. Age-related shifts in leaf chemistry of clonal aspen (*Populus tremuloides*). *Journal of Chemical Ecology*, 32(7):1415–1429, 2006.

[87] P.A. Duffy. *Interactions among climate, fire, and vegetation in the Alaskan boreal forest*. ProQuest, 2006.

[88] P.A. Duffy, J. Epting, J.M. Graham, S. Rupp, and D.A. McGuire. Analysis of Alaskan burn severity patterns using remotely sensed data. *International Journal of Wildland Fire*, 16(3):277–284, 2007.

[89] P.A. Duffy, J.E. Walsh, J.M. Graham, D.H. Mann, and S. Rupp. Impacts of large-scale atmospheric-ocean variability on Alaskan fire season severity. *Ecological Applications*, 15(4):1317–1330, 2005.

[90] J.R. Dugle. A taxonomic study of western Canadian species in the genus Betula. *Canadian Journal of Botany*, 44(7):929–1007, 1966.

[91] R. Dybzinski, C. Farrior, A. Wolf, P.B. Reich, and S.W. Pacala. Evolutionarily stable strategy carbon allocation to foliage, wood, and fine roots in trees competing for light and nitrogen: An analytically tractable, individual-based model and quantitative comparisons to data. *The American Naturalist*, 177(2):153–166, 2011.

[92] L. Edelstein-Keshet. Mathematical theory for plant-herbivore systems. *Journal of Mathematical Biology*, 24(1):25–58, 1986.

[93] L. Edelstein-Keshet and M.D. Rausher. The effects of inducible plant defenses on herbivore populations. 1. Mobile herbivores in continuous time. *The American Naturalist*, 133(6):787–810, 1989.

[94] M.G. Efford, D.L. Borchers, and A.E. Byrom. Density estimation by spatially explicit capture-recapture: likelihood-based methods. In *Modeling demographic processes in marked populations*, pages 255–269. Springer, 2009.

[95] Z. Feng, J.A. Alfaro-Murillo, D.L. DeAngelis, J. Schmidt, M. Barga, Y. Zheng, M.H.B.A. Tamrin, M. Olson, T. Glaser, K. Kielland, F.S. Chapin III, and J.P. Bryant. Plant toxins and trophic cascades alter fire regime and succession on a boreal forest landscape. *Ecological Modelling*, 244:79–92, 2012.

[96] Z. Feng, W. Huang, and D.L. DeAngelis. Spatially heterogeneous invasion of toxic plant mediated by herbivory. *Mathematical Biosciences & Engineering*, 10:1519–1538, 2013.

[97] Z. Feng, R. Liu, and D.L. DeAngelis. Plant-herbivore interactions mediated by plant toxicity. *Theoretical Population Biology*, 73(3):449–459, 2008.

[98] Z. Feng, R. Liu, D.L. DeAngelis, J.P. Bryant, K. Kielland, F.S. Chapin III, and R.K. Swihart. Plant toxicity, adaptive herbivory, and plant community dynamics. *Ecosystems*, 12(4):534–547, 2009.

[99] Z. Feng, Z. Qiu, R. Liu, and D.L. DeAngelis. Dynamics of a plant-herbivore-predator system with plant-toxicity. *Mathematical Biosciences*, 229(2):190–204, 2011.

[100] C. Field. Allocating leaf nitrogen for the maximization of carbon gain: leaf age as a control on the allocation program. *Oecologia*, 56(2-3):341–347, 1983.

[101] J.P. Finerty. *The population ecology of cycles in small mammals: mathematical theory and biological fact.* Yale University Press, 1980.

[102] N. Fisichelli, L.E. Frelich, and P.B. Reich. Sapling growth responses to warmer temperatures 'cooled' by browse pressure. *Global Change Biology*, 18(11):3455–3463, 2012.

[103] M. Flannigan, I. Campbell, M. Wotton, C. Carcaillet, P. Richard, and Y. Bergeron. Future fire in Canada's boreal forest: paleoecology results and general circulation model-regional climate model simulations. *Canadian Journal of Forest Research*, 31(5):854–864, 2001.

[104] J.L. Folse, Jane M. P., and W.E. Grant. AI modelling of animal movements in a heterogeneous habitat. *Ecological Modelling*, 46(1-2):57–72, 1989.

[105] J.S. Forbey, X. Pu, D. Xu, K. Kielland, and J.P. Bryant. Inhibition of snowshoe hare succinate dehydrogenase activity as a mechanism of deterrence for papyriferic acid in birch. *Journal of Chemical Ecology*, 37(12):1285–1293, 2011.

[106] D. Fortin, J.A. Merkle, M. Sigaud, S.G. Cherry, S. Plante, A. Drolet, and M. Labrecque. Temporal dynamics in the foraging decisions of large herbivores. *Animal Production Science*, 55(3):376–383, 2015.

[107] J.F. Fox. Forest fires and the snowshoe hare-Canada lynx cycle. *Oecologia*, 31(3):349–374, 1978.

[108] J.F. Fox and J.P. Bryant. Instability of the snowshoe hare and woody plant interaction. *Oecologia*, 63(1):128–135, 1984.

[109] L.R. Fox. Defense and dynamics in plant-herbivore systems. *American Zoologist*, 21(4):853–864, 1981.

[110] O. Franklin, J. Johansson, R.C. Dewar, U. Dieckmann, R.E. McMurtrie, Å. Brännström, and R. Dybzinski. Modeling carbon allocation in trees: a search for principles. *Tree Physiology*, 32(6):648–666, 2012.

[111] A.W. Franzmann and C.C. Schwartz. *Ecology and management of the North American moose.* University Press of Colorado, 2007.

[112] W.J. Freeland and D.H. Janzen. Strategies in herbivory by mammals: the role of plant secondary compounds. *The American Naturalist*, 108(961):269–289, 1974.

[113] S.D. Fretwell and J.S. Calver. On territorial behavior and other factors influencing habitat distribution in birds. *Acta Biotheoretica*, 19(1):37–44, 1969.

[114] A.L. Friend, M.D. Coleman, and J.G. Isebrands. Carbon allocation to root and shoot systems of woody plants. In *Biology of adventitious root formation*, pages 245–273. Springer, 1994.

[115] J.M. Fryxell. Forage quality and aggregation by large herbivores. *The American Naturalist*, 138(2):478–498, 1991.

[116] J.M. Fryxell. Evolutionary dynamics of habitat use. *Evolutionary Ecology*, 11(6):687–701, 1997.

[117] W.C. Gasaway, R.O. Stephenson, J.L. Davis, P.E.K. Shepherd, and O.E. Burris. Interrelationships of wolves, prey, and man in interior Alaska. *Wildlife Monographs*, (84):1–50, 1983.

[118] G.F. Gause. Experimental analysis of Vito Volterra's mathematical theory of the struggle for existence. *Science*, 79(2036):16–17, 1934.

[119] S. Gayler, T.E.E. Grams, W. Heller, D. Treutter, and E. Priesack. A dynamical model of environmental effects on allocation to carbon-based secondary compounds in juvenile trees. *Annals of Botany*, 101(8):1089–1098, 2008.

[120] S. Gayler and E. Priesack. PLATHO, a simulation model of resource allocation in the plant-soil system. Technical report, Institute of Soil Ecology GSF–National Research Center for Environment and Health (http://www.sfb607.de/english/projects/c2/platho.pdf), 2003.

[121] R.P. Gendron. Models and mechanisms of frequency-dependent predation. *The American Naturalist*, 130(4):603–623, 1987.

[122] S.A.H. Geritz, J.A.J. Metz, É. Kisdi, and G. Meszéna. Dynamics of adaptation and evolutionary branching. *Physical Review Letters*, 78(10):2024, 1997.

[123] J. Gershenzon. The cost of plant chemical defense against herbivory: a biochemeical perspective. In *Insect-plant interactions*, pages 105–173. CRC Press, 1994.

[124] J. Gershenzon. Metabolic costs of terpenoid accumulation in higher plants. *Journal of Chemical Ecology*, 20(6):1281–1328, 1994.

[125] J. Gershenzon and N. Dudareva. The function of terpene natural products in the natural world. *Nature Chemical Biology*, 3(7):408–414, 2007.

[126] J.F. Gilliam and D.F. Fraser. Habitat selection under predation hazard: test of a model with foraging minnows. *Ecology*, 68(6):1856–1862, 1987.

[127] T.J. Givnish. Plant stems: biomechanical adaptation for energy capture and influence on species. In *Plant stems: physiology and functional morphology*. Academic Press, 1995.

[128] S.K. Gleeson and D. Tilman. Plant allocation, growth rate and successional status. *Functional Ecology*, 8(4):543–550, 1994.

[129] J. Gorham, M. Tori, Y. Asakawa, et al. *The biochemistry of the stilbenoids*. Chapman & Hall, 1995.

[130] W.B. Grange. *Way to game abundance*. Scribner, 1949.

[131] W.B. Grange. Fire and tree growth relationships to snowshoe rabbits. In *Proceedings of the Tall Timbers Fire Ecology Conference*, volume 4, pages 110–125, 1965.

[132] T.R. Green and A.C. Ryan. Wound-induced proteinase inhibitor in plant leaves: a possible defense mechanism against insects. *Science*, 175(4023):776–777, 1972.

[133] J.E. Gross, L.A. Shipley, N.T. Hobbs, D.E. Spalinger, and B.A. Wunder. Functional response of herbivores in food-concentrated patches: tests of a mechanistic model. *Ecology*, 74(3):778–791, 1993.

[134] Y.L. Grossman and T.M. DeJong. Peach: a simulation model of reproductive and vegetative growth in peach trees. *Tree Physiology*, 14(4):329–345, 1994.

[135] J.P. Grover. Competition, herbivory, and enrichment: nutrient-based models for edible and inedible plants. *The American Naturalist*, 145(5):746–774, 1995.

[136] P.J. Grubb. A positive distrust in simplicity–lessons from plant defences and from competition among plants and among animals. *Journal of Ecology*, 80(4):585–610, 1992.

[137] N.G. Hairston, F.E. Smith, and L.B. Slobodkin. Community structure, population control, and competition. *The American Naturalist*, 94(879):421–425, 1960.

[138] S.R. Hall. Stoichiometrically explicit competition between grazers: species replacement, coexistence, and priority effects along resource supply gradients. *The American Naturalist*, 164(2):157–172, 2004.

[139] P.A. Hambäck, B.D. Inouye, P. Andersson, and N. Underwood. Effects of plant neighborhoods on plant-herbivore interactions: resource dilution and associational effects. *Ecology*, 95(5):1370–1383, 2014.

[140] J.L. Harper et al. Population biology of plants. *Population Biology of Plants*, 1977.

[141] E. Haukioja. On the role of plant defences in the fluctuation of herbivore populations. *Oikos*, 35(2):202–213, 1980.

[142] D.T. Haydon and A.L. Lloyd. On the origins of the Lotka–Volterra equations. *Bulletin of the Ecological Society of America*, 80(3):205–206, 1999.

[143] H.F. Heady. Palatability of herbage and animal preference. *Journal of Range Management*, 17(2):76–82, 1964.

[144] M.L. Heinselman. Fire intensity and frequency as factors in the distribution and structure of northern ecosystems [Canadian and Alaskan boreal forests, Rocky Mountain subalpine forests, Great Lakes-Acadian forests, includes history, management; Canada; USA]. Technical report, USDA Forest Service, 1981.

[145] M.R. Heithaus, A.J. Wirsing, D. Burkholder, J. Thomson, and L.M. Dill. Towards a predictive framework for predator risk effects: the interaction of landscape features and prey escape tactics. *Journal of Animal Ecology*, 78(3):556–562, 2009.

[146] C. Hély, Y. Bergeron, and M.D. Flannigan. Effects of stand composition on fire hazard in mixed-wood Canadian boreal forest. *Journal of Vegetation Science*, 11(6):813–824, 2000.

[147] G.M. Hengeveld, F. van Langevelde, T.A. Groen, and H.J. de Knegt. Optimal foraging for multiple resources in several food species. *The American Naturalist*, 174(1):102–110, 2009.

[148] D.A. Herbert, E.B. Rastetter, L. Gough, and G.R. Shaver. Species diversity across nutrient gradients: an analysis of resource competition in model ecosystems. *Ecosystems*, 7(3):296–310, 2004.

[149] C. Hermans, J.P. Hammond, P.J. White, and N. Verbruggen. How do plants respond to nutrient shortage by biomass allocation? *Trends in Plant Science*, 11(12):610–617, 2006.

[150] D.A. Herms. Effects of fertilization on insect resistance of woody ornamental plants: reassessing an entrenched paradigm. *Environmental Entomology*, 31(6):923–933, 2002.

[151] D.A. Herms and W.J. Mattson. The dilemma of plants: to grow or defend. *Quarterly Review of Biology*, pages 283–335, 1992.

[152] D.O. Hessen and E. Van Donk. Morphological changes in scenedesmus induced by substances released from daphnia. *Archiv Fur Hydrobiologie*, 127:129–129, 1993.

[153] D.S. Hik. Does risk of predation influence population dynamics? Evidence from cyclic decline of snowshoe hares. *Wildlife Research*, 22(1):115–129, 1995.

[154] D.W. Hilbert, D.M. Swift, J.K. Detling, and M.I. Dyer. Relative growth rates and the grazing optimization hypothesis. *Oecologia*, 51(1):14–18, 1981.

[155] N.T. Hobbs, J.E. Gross, L.A. Shipley, D.E. Spalinger, and B.A. Wunder. Herbivore functional response in heterogeneous environments: a contest among models. *Ecology*, 84(3):666–681, 2003.

[156] V. Hochman and B. Kotler. Effects of food quality, diet preference and water on patch use by Nubian ibex. *Oikos*, 112(3):547–554, 2006.

[157] J. Hodson, D. Fortin, and L. Bélanger. Fine-scale disturbances shape space-use patterns of a boreal forest herbivore. *Journal of Mammalogy*, 91(3):607–619, 2010.

[158] C.S. Holling. The components of predation as revealed by a study of small-mammal predation of the European pine sawfly. *The Canadian Entomologist*, 91(05):293–320, 1959.

[159] C.S. Holling. Some characteristics of simple types of predation and parasitism. *The Canadian Entomologist*, 91(07):385–398, 1959.

[160] R.D. Holt. Predation, apparent competition, and the structure of prey communities. *Theoretical Population Biology*, 12(2):197–229, 1977.

[161] A.I. Houston, A.D. Higginson, and J.M. McNamara. Optimal foraging for multiple nutrients in an unpredictable environment. *Ecology Letters*, 14(11):1101–1107, 2011.

[162] W. Huang. Traveling wave solutions for a class of predator-prey systems. *Journal of Dynamics and Differential Equations*, 24(3):633–644, 2012.

[163] M.R. Hutchings, I. Kyriazakis, T.G. Papachristou, I.J. Gordon, and F. Jackson. The herbivores' dilemma: trade-offs between nutrition and parasitism in foraging decisions. *Oecologia*, 124(2):242–251, 2000.

[164] N.J. Hutchings and I.J. Gordon. A dynamic model of herbivore-plant interactions on grasslands. *Ecological Modelling*, 136(2):209–222, 2001.

[165] A.R. Imre and J. Bogaert. The fractal dimension as a measure of the quality of habitats. *Acta Biotheoretica*, 52(1):41–56, 2004.

[166] T. Ingestad and G.I. Agren. The influence of plant nutrition on biomass allocation. *Ecological Applications*, 1(2):168–174, 1991.

[167] J.G. Irons III, J.P. Bryant, and M.W. Oswood. Effects of moose browsing on decomposition rates of birch leaf litter in a subarctic stream. *Canadian Journal of Fisheries and Aquatic Sciences*, 48(3):442–444, 1991.

[168] A.R. Ives and A.P. Dobson. Antipredator behavior and the population dynamics of simple predator-prey systems. *The American Naturalist*, 130(3):431–447, 1987.

[169] Y. Iwasa and D. Cohen. Optimal growth schedule of a perennial plant. *The American Naturalist*, 133(4):480–505, 1989.

[170] Y. Iwasa and J. Roughgarden. Shoot/root balance of plants: optimal growth of a system with many vegetative organs. *Theoretical Population Biology*, 25(1):78–105, 1984.

[171] W.J. Jakubas, W.H. Karasov, and C.G. Guglielmo. Ruffed grouse tolerance and biotransformation of the plant secondary metabolite coniferyl benzoate. *Condor*, pages 625–640, 1993.

[172] D.H. Janzen. Herbivores and the number of tree species in tropical forests. *The American Naturalist*, 104(940):501–528, 1970.

[173] J. Järemo, J. Tuomi, P. Nilsson, and T. Lennartsson. Plant adaptations to herbivory: mutualistic versus antagonistic coevolution. *Oikos*, 84(2):313–320, 1999.

[174] P. Jeandet, B. Delaunois, A. Conreux, D. Donnez, V. Nuzzo, S. Cordelier, C. Clément, and E. Courot. Biosynthesis, metabolism, molecular engineering, and biological functions of stilbene phytoalexins in plants. *Biofactors*, 36(5):331–341, 2010.

[175] J.M. Jeschke and R. Tollrian. Density-dependent effects of prey defences. *Oecologia*, 123(3):391–396, 2000.

[176] M.K. Jogia, A.R.E. Sinclair, and R.J. Andersen. An antifeedant in balsam poplar inhibits browsing by snowshoe hares. *Oecologia*, 79(2):189–192, 1989.

[177] J.F. Johnstone and F.S. Chapin III. Effects of soil burn severity on post-fire tree recruitment in boreal forest. *Ecosystems*, 9(1):14–31, 2006.

[178] J.F. Johnstone, S. Rupp, M. Olson, and D. Verbyla. Modeling impacts of fire severity on successional trajectories and future fire behavior in Alaskan boreal forests. *Landscape Ecology*, 26(4):487–500, 2011.

[179] S. Ju and D.L. DeAngelis. The R* rule and energy flux in a plant-nutrient ecosystem. *Journal of Theoretical Biology*, 256(3):326–332, 2009.

[180] R. Julkunen-Tiitto, M. Rousi, J. Bryant, S. Sorsa, M. Keinänen, and H. Sikanen. Chemical diversity of several betulaceae species: comparison of phenolics and terpenoids in northern birch stems. *Trees*, 11(1):16–22, 1996.

[181] E. Kaarlejärvi, R. Baxter, A. Hofgaard, H. Hytteborn, O. Khitun, U. Molau, S. Sjögersten, P. Wookey, and J. Olofsson. Effects of warming on shrub abundance and chemistry drive ecosystem-level changes in a forest-tundra ecotone. *Ecosystems*, 15(8):1219–1233, 2012.

[182] Y. Kang, D. Armbruster, and Y. Kuang. Dynamics of a plant-herbivore model. *Journal of Biological Dynamics*, 2(2):89–101, 2008.

[183] R. Karban and A.A. Agrawal. Herbivore offense. *Annual Review of Ecology and Systematics*, pages 641–664, 2002.

[184] R. Karban and I.T. Baldwin. *Induced responses to herbivory*. University of Chicago Press, 2007.

[185] L.B. Kats and L.M. Dill. The scent of death: chemosensory assessment of predation risk by prey animals. *Ecoscience*, 5(3):361–394, 1998.

[186] C.M.K. Kaunzinger and P.J. Morin. Productivity controls food-chain properties in microbial communities. *Nature*, 395(6701):495–497, 1998.

[187] L.B. Keith. Role of food in hare population cycles. *Oikos*, 40(3):385–395, 1983.

[188] L.B. Keith and L.A. Windberg. A demographic analysis of the snowshoe hare cycle. *Wildlife Monographs*, (58):3–70, 1978.

[189] J.G. Kie. Optimal foraging and risk of predation: effects on behavior and social structure in ungulates. *Journal of Mammalogy*, 80(4):1114–1129, 1999.

[190] K. Kielland and J.P. Bryant. Moose herbivory in taiga: effects on biogeochemistry and vegetation dynamics in primary succession. *Oikos*, 82(2):377–383, 1998.

[191] K. Kielland, J.P. Bryant, and R.W. Ruess. Moose herbivory and carbon turnover of early successional stands in interior Alaska. *Oikos*, 80(1):25–30, 1997.

[192] K. Kielland, J.P. Bryant, and R.W. Ruess. Mammalian herbivory, ecosystem engineering, and ecological cascades in Alaskan boreal forests. In *Alaska's changing boreal forest*, pages 211–226. Oxford University Press, New York, 2006.

[193] A.A. King and W.M. Schaffer. The geometry of a population cycle: a mechanistic model of snowshoe hare demography. *Ecology*, 82(3):814–830, 2001.

[194] D.A. King. A model analysis of the influence of root and foliage allocation on forest production and competition between trees. *Tree Physiology*, 12(2):119–135, 1993.

[195] D.A. King. Allocation of above-ground growth is related to light in temperate deciduous saplings. *Functional Ecology*, 17(4):482–488, 2003.

[196] T.T. Kozlowski. *Growth and development of trees. Volume II. Cambial growth, root growth, and reproduction growth*. Academic Press, 1971.

[197] P.J. Kramer and T.T. Kozlowski. *Physiology of woody plants*. Academic Press, New York, 1979.

[198] C. Krebs. Kluane monitoring. http://www.zoology.ubc.ca/~krebs/kluane.html.

[199] C.J. Krebs. Of lemmings and snowshoe hares: the ecology of northern Canada. *Proceedings of the Royal Society of London B: Biological Sciences*, 278(1705):481–489, 2011.

[200] C.J. Krebs, R. Boonstra, S. Boutin, and A.R.E. Sinclair. What drives the 10-year cycle of snowshoe hares? *BioScience*, 51(1):25–35, 2001.

[201] C.J. Krebs, S. Boutin, and R. Boonstra. *Ecosystem dynamics of the boreal forest–The Kluane Project*. Oxford University Press, 2001.

[202] C.J. Krebs, S. Boutin, R. Boonstra, A.R.E. Sinclair, et al. Impact of food and predation on the snowshoe hare cycle. *Science*, 269(5227):1112, 1995.

[203] L.W. Krefting and C.E. Ahlgren. Small mammals and vegetation changes after fire in a mixed conifer-hardwood forest. *Ecology*, 55(6):1391–1398, 1974.

[204] M. Kretzschmar, R.M. Nisbet, and E. McCauley. A predator-prey model for zooplankton grazing on competing algal populations. *Theoretical Population Biology*, 44(1):32–66, 1993.

[205] Y. Kuang, J. Huisman, J.J. Elser, et al. Stoichiometric plant-herbivore models and their interpretation. *Mathematical Biosciences and Engineering*, 1(2):215–222, 2004.

[206] H.W. Kuhlmann, J. Kusch, and K. Heckmann. Predator-induced defenses in ciliated protozoa. In *The ecology and evolution of inducible defenses*, pages 142–159. Princeton University Press, 1999.

[207] T.A. Kurkowski, D.H. Mann, S. Rupp, and D.L. Verbyla. Relative importance of different secondary successional pathways in an Alaskan boreal forest. *Canadian Journal of Forest Research*, 38(7):1911–1923, 2008.

[208] Y.A. Kuznetsov. *Elements of applied bifurcation theory*, volume 112. Springer Science & Business Media, 2013.

[209] W. Lampert, K.O. Rothhaupt, and E. Von Elert. Chemical induction of colony formation in a green alga (*Scenedesmus acutus*) by grazers (*Daphnia*). *Limnology and Oceanography*, 39(7):1543–1550, 1994.

[210] J.J. Landsberg. *Physiological ecology of forest production*. Academic Press, London, 1986.

[211] J.H. Langenheim. Higher plant terpenoids: a phytocentric overview of their ecological roles. *Journal of Chemical Ecology*, 20(6):1223–1280, 1994.

[212] X. Le Roux, A. Lacointe, A. Escobar-Gutiérrez, and S. Le Dizès. Carbon-based models of individual tree growth: a critical appraisal. *Annals of Forest Science*, 58(5):469–506, 2001.

[213] M.A. Leibold. Resource edibility and the effects of predators and productivity on the outcome of trophic interactions. *The American Naturalist*, 134(6):922–949, 1989.

[214] M.A. Leibold. A graphical model of keystone predators in food webs: trophic regulation of abundance, incidence, and diversity patterns in communities. *The American Naturalist*, 147(5):784–812, 1996.

[215] M.A. Leibold, J.M. Chase, J.B. Shurin, and A.L. Downing. Species turnover and the regulation of trophic structure. *Annual Review of Ecology and Systematics*, 28(1):467–494, 1997.

[216] A. Leopold. The conservation ethic. *Journal of Forestry*, 31(6):634–643, 1933.

[217] B-L. Li. Fractal geometry applications in description and analysis of patch patterns and patch dynamics. *Ecological Modelling*, 132(1):33–50, 2000.

[218] Y. Li and Z. Feng. Dynamics of a plant-herbivore model with toxin-induced functional response. *Mathematical Biosciences and Engineering*, 7(1):149–169, 2010.

[219] S.L. Lima. Nonlethal effects in the ecology of predator-prey interactions. *BioScience*, 48(1):25–34, 1998.

[220] L.E. Lindberg, S.M. Willför, and B.R. Holmbom. Antibacterial effects of knotwood extractives on paper mill bacteria. *Journal of Industrial Microbiology and Biotechnology*, 31(3):137–147, 2004.

[221] R.L. Lindeman. The trophic-dynamic aspect of ecology. *Ecology*, 23(4):399–417, 1942.

[222] C.M. Litton, J.W. Raich, and M.G. Ryan. Carbon allocation in forest ecosystems. *Global Change Biology*, 13(10):2089–2109, 2007.

[223] J.A. Litvaitis, J.A. Sherburne, and J.A. Bissonette. Influence of understory characteristics on snowshoe hare habitat use and density. *The Journal of Wildlife Management*, 49(4):866–873, 1985.

[224] R. Liu, D.L. DeAngelis, and J.P. Bryant. Dynamics of herbivores and resources on a landscape with interspersed resources and refuges. *Theoretical Ecology*, 7(2):195–208, 2014.

[225] R. Liu, D.L. DeAngelis, and J.P. Bryant. Ratio-dependent functional response emerges from optimal foraging on a complex landscape. *Ecological Modelling*, 292:45–50, 2014.

[226] R. Liu, Z. Feng, H. Zhu, and D.L. DeAngelis. Bifurcation analysis of a plant–herbivore model with toxin-determined functional response. *Journal of Differential Equations*, 245(2):442–467, 2008.

[227] R. Liu, S.A. Gourley, D.L. DeAngelis, and J.P. Bryant. Modeling the dynamics of woody plant-herbivore interactions with age-dependent toxicity. *Journal of Mathematical Biology*, 65(3):521–552, 2012.

[228] J.A. Logan, P. White, B.J. Bentz, and J.A. Powell. Model analysis of spatial patterns in mountain pine beetle outbreaks. *Theoretical Population Biology*, 53(3):236–255, 1998.

[229] I. Loladze, Y. Kuang, and J.J. Elser. Stoichiometry in producer-grazer systems: linking energy flow with element cycling. *Bulletin of Mathematical Biology*, 62(6):1137–1162, 2000.

[230] A.J. Lotka. *Elements of physical biology*. Williams and Wilkins Company, 1925.

[231] J. Luan, R.I. Muetzelfeldt, and J. Grace. Hierarchical approach to forest ecosystem simulation. *Ecological Modelling*, 86(1):37–50, 1996.

[232] D. Ludwig, D.D. Jones, and C.S. Holling. Qualitative analysis of insect outbreak systems: the spruce budworm and forest. *The Journal of Animal Ecology*, 47(1):315–332, 1978.

[233] P. Lundberg. Functional response of a small mammalian herbivore: the disc equation revisited. *The Journal of Animal Ecology*, 57(3):999–1006, 1988.

[234] P. Lundberg and M. Astrom. Low nutritive quality as a defense against optimally foraging herbivores. *The American Naturalist*, 135(4):547–562, 1990.

[235] H. Lyr. Detoxification of heartwood toxins and chlorophenols by higher fungi. *Nature*, 195(4838):289–290, 1962.

[236] R.H. MacArthur and E.R. Pianka. On optimal use of a patchy environment. *The American Naturalist*, 100(916):603–609, 1966.

[237] J.A.K. Maier, J.M. Ver Hoef, D.A. McGuire, T.R. Bowyer, L. Saperstein, and H.A. Maier. Distribution and density of moose in relation to landscape characteristics: effects of scale. *Canadian Journal of Forest Research*, 35(9):2233–2243, 2005.

[238] A. Mäkelä, H.T. Valentine, and H. Helmisaari. Optimal co-allocation of carbon and nitrogen in a forest stand at steady state. *New Phytologist*, 180(1):114–123, 2008.

[239] J. Mallet. The struggle for existence: how the notion of carrying capacity, K, obscures the links between demography, Darwinian evolution, and speciation. *Evolutionary Ecology Research*, 14(5):627–665, 2012.

[240] K.J. Marsh, I.R. Wallis, and W.J. Foley. Detoxification rates constrain feeding in common brushtail possums (*Trichosurus vulpecula*). *Ecology*, 86(11):2946–2954, 2005.

[241] K.J. Marsh, I.R. Wallis, and W.J. Foley. Behavioural contributions to the regulated intake of plant secondary metabolites in koalas. *Oecologia*, 154(2):283–290, 2007.

[242] K.J. Marsh, I.R. Wallis, S. McLean, J.S. Sorensen, and W.J. Foley. Conflicting demands on detoxification pathways influence how common brushtail possums choose their diets. *Ecology*, 87(8):2103–2112, 2006.

[243] R.S. McAlpine and B.M. Wotton. The use of fractal dimension to improve wildland fire perimeter predictions. *Canadian Journal of Forest Research*, 23(6):1073–1077, 1993.

[244] R.J.P. McAvinchey. *Winter herbivory by snowshoe hares and moose as a process affecting primary succession on an Alaskan floodplain.* 1991.

[245] K.D.M. McConnaughay and J.S. Coleman. Biomass allocation in plants: ontogeny or optimality? A test along three resource gradients. *Ecology*, 80(8):2581–2593, 1999.

[246] P.F. McInnes, R.J. Naiman, J. Pastor, and Y. Cohen. Effects of moose browsing on vegetation and litter of the boreal forest, Isle Royale, Michigan, USA. *Ecology*, 73(6):2059–2075, 1992.

[247] Doyle McKey. The distribution of secondary compounds within plants. In *Herbivores: their interaction with secondary plant metabolites*, pages 55–133. Academic Press, New York, 1979.

[248] S. McLean and A.J. Duncan. Pharmacological perspectives on the detoxification of plant secondary metabolites: implications for ingestive behavior of herbivores. *Journal of Chemical Ecology*, 32(6):1213–1228, 2006.

[249] S. McLean, S.M. Richards, S. Cover, S. Brandon, N.W. Davies, J.P. Bryant, and T.P. Clausen. Papyriferic acid, an antifeedant triterpene from birch trees, inhibits succinate dehydrogenase from liver mitochondria. *Journal of Chemical Ecology*, 35(10):1252, 2009.

[250] J.M. McNamara and A.I. Houston. The common currency for behavioral decisions. *The American Naturalist*, 127(3):358–378, 1986.

[251] J.M. McNamara and A.I. Houston. Starvation and predation as factors limiting population size. *Ecology*, 68(5):1515–1519, 1987.

[252] J.M. McNamara and A.I. Houston. Risk-sensitive foraging: a review of the theory. *Bulletin of Mathematical Biology*, 54(2-3):355–378, 1992.

[253] S.J. McNaughton. Grazing as an optimization process: grass-ungulate relationships in the Serengeti. *The American Naturalist*, 113(5):691–703, 1979.

[254] M.A. Melchiors and C.A. Leslie. Effectiveness of predator fecal odors as black-tailed deer repellents. *The Journal of Wildlife Management*, 49(2):358–362, 1985.

[255] M.C. Melnychuk and C.J. Krebs. Residual effects of NPK fertilization on shrub growth in a Yukon boreal forest. *Canadian Journal of Botany*, 83(4):399–404, 2005.

[256] R. Moen, Y. Cohen, and J. Pastor. Evaluating foraging strategies with a moose energetics model. *Ecosystems*, 1:52–63, 1998.

[257] R. Moen, J. Pastor, and Y. Cohen. A spatially explicit model of moose foraging and energetics. *Ecology*, 78(2):505–521, 1997.

[258] J. Moorby and P.F. Wareing. Ageing in woody plants. *Annals of Botany*, 27(2):291–308, 1963.

[259] M.A. Mouissie, E.F. Apol, G.W. Heil, and R. van Diggelen. Creation and preservation of vegetation patterns by grazing. *Ecological Modelling*, 218(1):60–72, 2008.

[260] D. Müller-Schwarze, H. Brashear, R. Kinnel, K.A. Hintz, A. Lioubomirov, and C. Skibo. Food processing by animals: do beavers leach tree bark to improve palatability? *Journal of Chemical Ecology*, 27(5):1011–1028, 2001.

[261] W.W. Murdoch. Community structure, population control, and competition–A critique. *The American Naturalist*, 100(912):219–226, 1966.

[262] J.H. Myers and B.J. Campbell. Distribution and dispersal in populations capable of resource depletion. *Oecologia*, 24(1):7–20, 1976.

[263] I.H. Myers-Smith, B.C. Forbes, M. Wilmking, M. Hallinger, T. Lantz, D. Blok, K.D. Tape, M. Macias-Fauria, U. Sass-Klaassen, E. Lévesque, et al. Shrub expansion in tundra ecosystems: dynamics, impacts and research priorities. *Environmental Research Letters*, 6(4):045509, 2011.

[264] R. Nathan and R. Casagrandi. A simple mechanistic model of seed dispersal, predation and plant establishment: Janzen-Connell and beyond. *Journal of Ecology*, 92(5):733–746, 2004.

[265] R.P. Neilson. Transient ecotone response to climatic change: some conceptual and modelling approaches. *Ecological Applications*, 3(3):385–395, 1993.

[266] J.L. Nelson, E.S. Zavaleta, and F. S. Chapin III. Boreal fire effects on subsistence resources in Alaska and adjacent Canada. *Ecosystems*, 11(1):156–171, 2008.

[267] A.J. Nicholson and V.A. Bailey. The balance of animal populations–Part I. In *Proceedings of the Zoological Society of London*, volume 105, pages 551–598. Wiley Online Library, 1935.

[268] I. Noy-Meir. Stability of grazing systems: an application of predator-prey graphs. *The Journal of Ecology*, 63(2):459–481, 1975.

[269] A. Oaten and W.W. Murdoch. Switching, functional response, and stability in predator-prey systems. *The American Naturalist*, 109(967):299–318, 1975.

[270] Flora of North America online 1. Distribution map, taxon: *Betula neoalaskana*. http://www.efloras.org/object_page.aspx?object_id=5748&flora_id=1, (accessed January 7, 2017).

[271] Flora of North America online 2. Distribution map, taxon: *Betula papyrifera*. http://www.efloras.org/object_page.aspx?object_id=5753&flora_id=1, (accessed January 7, 2017).

[272] Flora of North America online 3. Flora of north america, **betula neoalaskana** sargent. http://www.efloras.org/florataxon.aspx?flora_id=1&taxon_id=233500257, (accessed January 7, 2017).

[273] Flora of North America online 4. Flora of north america, **betula papyrifera** marshall. http://www.efloras.org/florataxon.aspx?flora_id=1&taxon_id=233500260, (accessed January 7, 2017).

[274] L. Oksanen, S.D. Fretwell, J. Arruda, and P. Niemela. Exploitation ecosystems in gradients of primary productivity. *The American Naturalist*, 118(2):240–261, 1981.

[275] L. Oksanen and P. Lundberg. Optimization of reproductive effort and foraging time in mammals: the influence of resource level and predation risk. *Evolutionary Ecology*, 9(1):45–56, 1995.

[276] L. Oksanen and T. Oksanen. The logic and realism of the hypothesis of exploitation ecosystems. *The American Naturalist*, 155(6):703–723, 2000.

[277] J. Olofsson, L. Oksanen, T. Callaghan, P.E. Hulme, T. Oksanen, and O. Suominen. Herbivores inhibit climate-driven shrub expansion on the tundra. *Global Change Biology*, 15(11):2681–2693, 2009.

[278] M.W. Oswood, N.F. Hughes, and A.M. Milner. Running waters of the Alaskan boreal forest. In *Alaska's changing boreal forest*, pages 147–170. Oxford University Press, New York, 2006.

[279] N. Owen-Smith and P. Novellie. What should a clever ungulate eat? *The American Naturalist*, 119(2):151–178, 1982.

[280] N.R. Owen-Smith. *Adaptive herbivore ecology: from resources to populations in variable environments*. Cambridge University Press, 2002.

[281] R.T. Paine. Phycology for the mammalogist: marine rocky shores and mammal-dominated communities – how different are the structuring processes? *Journal of Mammalogy*, 81(3):637–648, 2000.

[282] T.R. Palo and C.T. Robbins. *Plant defenses against mammalian herbivory*. CRC Press, 1991.

[283] J.M. Paruelo, S. Pütz, G. Weber, M. Bertiller, R.A. Golluscio, M.R. Aguiar, and T. Wiegand. Long-term dynamics of a semiarid grass steppe under stochastic climate and different grazing regimes: A simulation analysis. *Journal of Arid Environments*, 72(12):2211–2231, 2008.

[284] J. Pastor. *What should a clever moose eat?* Island Press, 2016.

[285] J. Pastor, Y. Cohen, and R. Moen. Generation of spatial patterns in boreal forest landscapes. *Ecosystems*, 2(5):439–450, 1999.

[286] J. Pastor, B. Dewey, R. Moen, D.J. Mladenoff, M. White, and Y. Cohen. Spatial patterns in the moose-forest-soil ecosystem on Isle Royale, Michigan, USA. *Ecological Applications*, 8(2):411–424, 1998.

[287] J. Pastor and R.J. Naiman. Selective foraging and ecosystem processes in boreal forests. *The American Naturalist*, 139(4):690–705, 1992.

[288] J. Pastor, R.J. Naiman, B. Dewey, and P. McInnes. Moose, microbes, and the boreal forest. *BioScience*, 38(11):770–777, 1988.

[289] R. Pearl and L.J. Reed. On the rate of growth of the population of the United States since 1790 and its mathematical representation. *Proceedings of the National Academy of Sciences*, 6(6):275–288, 1920.

[290] J.L. Pease, R.H. Vowles, and L.B. Keith. Interaction of snowshoe hares and woody vegetation. *The Journal of Wildlife Management*, 43(1):43–60, 1979.

[291] D.A. Perry. *Forest ecosystems*. JHU Press, 1994.

[292] S.L. Pimm. Food webs. In *Food webs*, pages 1–11. Springer, 1982.

[293] R.S. Poethig. The past, present, and future of vegetative phase change. *Plant Physiology*, 154(2):541–544, 2010.

[294] E. Post and C. Pedersen. Opposing plant community responses to warming with and without herbivores. *Proceedings of the National Academy of Sciences*, 105(34):12353–12358, 2008.

[295] M.E. Power. Top-down and bottom-up forces in food webs: do plants have primacy. *Ecology*, 73(3):733–746, 1992.

[296] M.V. Price and J.W. Joyner. What resources are available to desert granivores: seed rain or soil seed bank? *Ecology*, 78(3):764–773, 1997.

[297] P.W. Price, C.E. Bouton, P. Gross, B.A. McPheron, J.N. Thompson, and A.E. Weis. Interactions among three trophic levels: influence of plants on interactions between insect herbivores and natural enemies. *Annual Review of Ecology and Systematics*, 11:41–65, 1980.

[298] F.D. Provenza, J.J. Villalba, L.E. Dziba, S.B. Atwood, and R.E. Banner. Linking herbivore experience, varied diets, and plant biochemical diversity. *Small Ruminant Research*, 49(3):257–274, 2003.

[299] F.D. Provenza, J.J. Villalba, J. Haskell, J.W. MacAdam, T.C. Griggs, and R.D. Wiedmeier. The value to herbivores of plant physical and chemical diversity in time and space. *Crop Science*, 47(1):382–398, 2007.

[300] R.H. Pulliam. On the theory of optimal diets. *The American Naturalist*, 108(959):59–74, 1974.

[301] E.B. Rastetter, M.G. Ryan, G.R. Shaver, J.M. Melillo, K.J. Nadelhoffer, J.E. Hobbie, and J.D. Aber. A general biogeochemical model describing the responses of the C and N cycles in terrestrial ecosystems to changes in CO2, climate, and N deposition. *Tree Physiology*, 9(1-2):101–126, 1991.

[302] H.M. Rauscher, J.G. Isebrands, G.E. Host, R.E. Dickson, D.I. Dickmann, T.R. Crow, and D.A. Michael. ECOPHYS: an ecophysiological growth process model for juvenile poplar. *Tree Physiology*, 7(1-2-3-4):255–281, 1990.

[303] L.A. Real and S.A. Levin. Theoretical advances: the role of theory in the rise of modern ecology. In *Foundations of ecology: Classic papers with commentaries*, pages 177–191. The University of Chicago Press, 1991.

[304] P.B. Reichardt. Papyriferic acid: a triterpenoid from Alaskan paper birch. *The Journal of Organic Chemistry*, 46(22):4576–4578, 1981.

[305] P.B. Reichardt. The chemistry of plant/animal interactions. In *National Wildlife Research Center Repellents Conference 1995*, page 30, 1995.

[306] P.B. Reichardt, J.P. Bryant, B.J. Anderson, D. Phillips, T.P. Clausen, M. Meyer, and K. Frisby. Germacrone defends labrador tea from browsing by snowshoe hares. *Journal of Chemical Ecology*, 16(6):1961–1970, 1990.

[307] P.B. Reichardt, J.P. Bryant, T.P. Clausen, and G.D. Wieland. Defense of winter-dormant Alaska paper birch against snowshoe hares. *Oecologia*, 65(1):58–69, 1984.

[308] P.B. Reichardt, J.P. Bryant, B.R. Mattes, T.P. Clausen, F.S. Chapin III, and M. Meyer. Winter chemical defense of Alaskan balsam poplar against snowshoe hares. *Journal of Chemical Ecology*, 16(6):1941–1959, 1990.

[309] P.B. Reichardt, T.P. Green, and S. Chang. 3-O-malonylbetulafolientriol oxide I from *Betula nana* subsp. *exilis*. *Phytochemistry*, 26(3):855–856, 1987.

[310] O.J. Reichman and M.V. Price. Ecological aspects of heteromyid foraging. In *Biology of the Heteromyidae*, number 10, pages 539–574. American Society of Mammalogists, 1993.

[311] D.F. Rhoades. Evolution of plant chemical defense against herbivores. In *Herbivores: their interaction with secondary plant metabolites*, pages 3–54. Academic Press, New York, 1979.

[312] D.F. Rhoades. Offensive-defensive interactions between herbivores and plants: their relevance in herbivore population dynamics and ecological theory. *The American Naturalist*, 125(2):205–238, 1985.

[313] K.L. Risenhoover, L.A. Renecker, and L.E. Morgantini. Effects of secondary metabolites from balsam poplar and paper birch on cellulose digestion. *Journal of Range Management*, 38(4):370–372, 1985.

[314] A.R. Rodgers and A.R.E. Sinclair. Diet choice and nutrition of captive snowshoe hares (*lepus americanus*): interactions of energy, protein, and plant secondary compounds. *Ecoscience*, 4(2):163–169, 1997.

[315] G. Rosenthal and M. Berenbaum. *Herbivores: their interaction with plant secondary metabolites*. Academic Press, New York, 1992.

[316] M.L. Rosenzweig et al. Paradox of enrichment: destabilization of exploitation ecosystems in ecological time. *Science*, 171(3969):385–387, 1971.

[317] M.L. Rosenzweig and R.H. MacArthur. Graphical representation and stability conditions of predator-prey interactions. *The American Naturalist*, 97(895):209–223, 1963.

[318] J.W. Rowe. *Natural products of woody plants: chemicals extraneous to the lignocellulosic cell wall*. Springer Science & Business Media, 2012.

[319] S. Rupp, F.S. Chapin III, and A.M. Starfield. Response of subarctic vegetation to transient climatic change on the Seward Peninsula in north-west Alaska. *Global Change Biology*, 6(5):541–555, 2000.

[320] S. Rupp, A.M. Starfield, and F.S. Chapin III. A frame-based spatially explicit model of subarctic vegetation response to climatic change: comparison with a point model. *Landscape Ecology*, 15(4):383–400, 2000.

[321] S. Rupp, A.M. Starfield, F.S. Chapin III, and P. Duffy. Modeling the impact of black spruce on the fire regime of Alaskan boreal forest. *Climatic Change*, 55(1-2):213–233, 2002.

[322] M. Scheffer. *Critical transitions in nature and society*. Princeton University Press, 2009.

[323] F.D. Schneider and S. Kéfi. Spatially heterogeneous pressure raises risk of catastrophic shifts. *Theoretical Ecology*, 9(2):207–217, 2016.

[324] C.C. Schwartz, W.L. Regelin, and A.W. Franzmann. Estimates of digestibility of birch, willow, and aspen mixtures in moose. *The Journal of Wildlife Management*, 52(1):33–37, 1988.

[325] K.R. Searle, N.T. Hobbs, and L.A. Shipley. Should I stay or should I go? Patch departure decisions by herbivores at multiple scales. *Oikos*, 111(3):417–424, 2005.

[326] D.S. Seigler. *Plant secondary metabolism*. Springer Science & Business Media, 2012.

[327] R.L. Senft, M.B. Coughenour, D.W. Bailey, L.R. Rittenhouse, O.E. Sala, and D.M. Swift. Large herbivore foraging and ecological hierarchies. *BioScience*, 37(11):789–799, 1987.

[328] G.G. Shaw, C.H.A. Little, and D.J. Durzan. Effect of fertilization of balsam fir trees on spruce budworm nutrition and development. *Canadian Journal of Forest Research*, 8(4):364–374, 1978.

[329] M.J. Sheriff, C.J. Krebs, and R. Boonstra. The ghosts of predators past: population cycles and the role of maternal programming under fluctuating predation risk. *Ecology*, 91(10):2983–2994, 2010.

[330] M.J. Sheriff, C.J. Krebs, and R. Boonstra. From process to pattern: how fluctuating predation risk impacts the stress axis of snowshoe hares during the 10-year cycle. *Oecologia*, 166(3):593–605, 2011.

[331] A. Sih. Prey uncertainty and the balancing of antipredator and feeding needs. *The American Naturalist*, 139(5):1052–1069, 1992.

[332] A. Sih and B. Christensen. Optimal diet theory: when does it work, and when and why does it fail? *Animal Behaviour*, 61(2):379–390, 2001.

[333] A.R.E. Sinclair, M.K. Jogia, and R.J. Andersen. Camphor from juvenile white spruce as an antifeedant for snowshoe hares. *Journal of Chemical Ecology*, 14(6):1505–1514, 1988.

[334] A.R.E. Sinclair, D.E. Williams, and R.J. Andersen. Triterpene constituents of the dwarf birch, *Betula glandulosa*. *Phytochemistry*, 31(7):2321–2324, 1992.

[335] J.N.M. Smith, C.J. Krebs, A.R.E. Sinclair, and R. Boonstra. Population biology of snowshoe hares. II. Interactions with winter food plants. *The Journal of Animal Ecology*, 57(1):269–286, 1988.

[336] M.E. Solomon. The natural control of animal populations. *The Journal of Animal Ecology*, 18(1):1–35, 1949.

[337] J.S. Sorensen, E. Heward, and M.D. Dearing. Plant secondary metabolites alter the feeding patterns of a mammalian herbivore (*Neotoma lepida*). *Oecologia*, 146(3):415–422, 2005.

[338] D.E. Spalinger and N.T. Hobbs. Mechanisms of foraging in mammalian herbivores: new models of functional response. *The American Naturalist*, 140(2):325–348, 1992.

[339] J.D.M. Speed, G. Austrheim, A.J. Hester, and A. Mysterud. Elevational advance of alpine plant communities is buffered by herbivory. *Journal of Vegetation Science*, 23(4):617–625, 2012.

[340] D.L. Spencer and J.B. Hakala. Moose and fire on the Kenai. In *Tall timbers fire ecology conference*, volume 3, pages 11–33, 1964.

[341] N. Stamp. Out of the quagmire of plant defense hypotheses. *The Quarterly Review of Biology*, 78(1):23–55, 2003.

[342] C.F. Steiner. The effects of prey heterogeneity and consumer identity on the limitation of trophic-level biomass. *Ecology*, 82(9):2495–2506, 2001.

[343] D.W. Stephens and J.R. Krebs. *Foraging theory*. Princeton University Press, 1986.

[344] M.T. Stevens, S.C. Brown, H.M. Bothwell, and J.P. Bryant. Biogeography of Alaska paper birch (*Betula neoalaskana*): latitudinal patterns in chemical defense and plant architecture. *Ecology*, 97(2):494–502, 2016.

[345] B.J. Stocks, J.A. Mason, J.B. Todd, E.M. Bosch, B.M. Wotton, B.D. Amiro, M.D. Flannigan, K.G. Hirsch, K.A. Logan, D.L. Martell, et al. Large forest fires in Canada, 1959–1997. *Journal of Geophysical Research: Atmospheres*, 107(D1), 2002.

[346] L.A. Stoddart and A.D. Smith. *Range management*. McGraw-Hill, 1955.

[347] T.P. Sullivan, D.R. Crump, H. Wieser, and E.A. Dixon. Influence of the plant antifeedant, pinosylvin, on suppression of feeding by snowshoe hares. *Journal of Chemical Ecology*, 18(7):1151–1164, 1992.

[348] R.K. Swihart, D.L. DeAngelis, Z. Feng, and J.P. Bryant. Troublesome toxins: time to re-think plant-herbivore interactions in vertebrate ecology. *BMC Ecology*, 9(1):1, 2009.

[349] H.T. Taipale, L. Härmälä, M. Rousi, and S.P. Lapinjoki. Histological and chemical comparison of triterpene and phenolic deterrent contents of juvenile shoots of *Betula* species. *Trees*, 8(5):232–236, 1994.

[350] K.W. Tang. Grazing and colony size development in *phaeocystis globosa* (prymnesiophyceae): the role of a chemical signal. *Journal of Plankton Research*, 25(7):831–842, 2003.

[351] J. Terborgh, K. Feeley, M. Silman, P. Nuñez, and B. Balukjian. Vegetation dynamics of predator-free land-bridge islands. *Journal of Ecology*, 94(2):253–263, 2006.

[352] C.A. Thanos. Aristotle and Theophrastus on plant-animal interactions. In *Plant-animal interactions in Mediterranean-type ecosystems*, pages 3–11. Springer, 1994.

[353] H.R. Thieme. Convergence results and a Poincaré–Bendixson trichotomy for asymptotically autonomous differential equations. *Journal of Mathematical Biology*, 30(7):755–763, 1992.

[354] J.H.M. Thornley. A model to describe the partitioning of photosynthate during vegetative plant growth. *Annals of Botany*, 36(2):419–430, 1972.

[355] J.H.M. Thornley. Shoot: root allocation with respect to C, N and P: an investigation and comparison of resistance and teleonomic models. *Annals of Botany*, 75(4):391–405, 1995.

[356] J.H.M. Thornley and I.R. Johnson. *Plant and crop modeling – A mathematical approach to plant and crop physiology*. The Blackburn Press, 1990.

[357] D. Tilman. *Resource competition and community structure*. Princeton University Press, 1982.

[358] L. Tinbergen. The natural control of insects in pinewoods. *Archives neerlandaises de zoologie*, 13(3):265–343, 1960.

[359] R. Tollrian and D.C. Harvell. *The ecology and evolution of inducible defenses*. Princeton University Press, 1999.

[360] J. Tuomi. Toward integration of plant defence theories. *Trends in Ecology & Evolution*, 7(11):365–367, 1992.

[361] P. Turchin. *Complex population dynamics: a theoretical/empirical synthesis*, volume 35. Princeton University Press, 2003.

[362] R. Turkington, E. John, C.J. Krebs, M.R.T. Dale, V.O. Nams, R. Boonstra, S. Boutin, K. Martin, A.R.E. Sinclair, and J. Smith. The effects of NPK fertilization for nine years on boreal forest vegetation in northwestern Canada. *Journal of Vegetation Science*, 9(3):333–346, 1998.

[363] K. Van Cleve and L.A. Viereck. A comparison of successional sequences following fire on permafrost-dominated and permafrost-free sites in interior Alaska. In *Permafrost: Fourth International Conference, Proceedings*, pages 1286–1291, 1983.

[364] J. van de Koppel and H.H.T. Prins. The importance of herbivore interactions for the dynamics of African savanna woodlands: an hypothesis. *Journal of Tropical Ecology*, 14(5):565–576, 1998.

[365] E. van Donk, M. Lürling, and W. Lampert. Consumer-induced changes in phytoplankton: inducibility, costs, benefits and the impact on grazers. In *The ecology and evolution of inducible defenses*, pages 89–103. Princeton University Press, 1999.

[366] H. van Oene, E.J.M. van Deursen, and F. Berendse. Plant-herbivore interaction and its consequences for succession in wetland ecosystems: a modeling approach. *Ecosystems*, 2(2):122–138, 1999.

[367] M.R. Vaughan and L.B. Keith. Demographic response of experimental snowshoe hare populations to overwinter food shortage. *The Journal of Wildlife Management*, 45(2):354–380, 1981.

[368] J.J. Villalba, F.D. Provenza, and J.P. Bryant. Consequences of nutrient-toxin inter-actions for herbivore selectivity: benefits or detriments for plants. *Oikos*, 97:282–292, 2002.

[369] J.J. Villalba, F.D. Provenza, and R. Shaw. Sheep self-medicate when challenged with illness-inducing foods. *Animal Behaviour*, 71(5):1131–1139, 2006.

[370] V. Volterra. Fluctuations in the abundance of a species considered mathematically. *Nature*, 118:558–560, 1926.

[371] V. Volterra. Variations and fluctuations of the number of individuals in animal species living together. *J. Cons. Int. Explor. Mer*, 3(1):3–51, 1928.

[372] M. Vos, B.W. Kooi, D.L. DeAngelis, and W.M. Mooij. Inducible defences and the paradox of enrichment. *Oikos*, 105(3):471–480, 2004.

[373] M. Vos, A.M. Verschoor, B.W. Kooi, F.L. Wäckers, D.L. DeAngelis, and W.M. Mooij. Inducible defenses and trophic structure. *Ecology*, 85(10):2783–2794, 2004.

[374] M. Vos, A.M. Verschoor, B.W. Kooi, F.L. Wäckers, D.L. DeAngelis, and W.M. Mooij. Inducible defenses and trophic structure. *Ecology*, 85(10):2783–2794, 2004.

[375] J.A. Vucetich and R.O. Peterson. The influence of top-down, bottom-up and abiotic factors on the moose (*Alces alces*) population of Isle Royale. *Proceedings of the Royal Society of London B: Biological Sciences*, 271(1535):183–189, 2004.

[376] D.J. Walker and N.C. Kenkel. Fractal analysis of spatio-temporal dynamics in boreal forest landscapes. *Abstracta Botanica*, 22:13–28, 1998.

[377] G.R. Walther, E. Post, P. Convey, A. Menzel, C. Parmesan, T.J.C. Beebee, J. Fro-mentin, O. Hoegh-Guldberg, and F. Bairlein. Ecological responses to recent climate change. *Nature*, 416(6879):389–395, 2002.

[378] D.A. Weinstein, R.D. Yanai, R. Beloin, and C.G. Zollweg. *The response of plants to interacting stresses: TREGRO Version 1.74–description and parameter requirements*. Electric Power Res. Institute, Palo Alto, 1992.

[379] B. Wermelinger, J. Baumgärtner, and A.P. Gutierrez. A demographic model of as-similation and allocation of carbon and nitrogen in grapevines. *Ecological Modelling*, 53:1–26, 1991.

[380] E.E. Werner and B.R. Anholt. Ecological consequences of the trade-off between growth and mortality rates mediated by foraging activity. *The American Natural-ist*, 142(2):242–272, 1993.

[381] M. Westoby. An analysis of diet selection by large generalist herbivores. *The American Naturalist*, 108(961):290–304, 1974.

[382] M. Westoby. What are the biological bases of varied diets? *The American Naturalist*, 112(985):627–631, 1978.

[383] T.C.R. White. *Why does the world stay green?: Nutrition and survival of plant-eaters*. CSIRO, 2005.

[384] T.C.R. White. *The inadequate environment: nitrogen and the abundance of animals*. Springer Science & Business Media, 2012.

[385] J.W. Williams, B.N. Shuman, T. Webb, P.J. Bartlein, and P.L. Leduc. Late-quaternary vegetation dynamics in North America: scaling from taxa to biomes. *Ecological Monographs*, 74(2):309–334, 2004.

[386] J.B. Wilson. A review of evidence on the control of shoot: root ratio, in relation to models. *Annals of Botany*, 61(4):433–449, 1988.

[387] J.O. Wolff. The role of habitat patchiness in the population dynamics of snowshoe hares. *Ecological Monographs*, 50(1):111–130, 1980.

[388] Q. Yu, H.E. Epstein, and D.A. Walker. Modeling dynamics of tundra plant communities on the Yamal Peninsula, Russia. In *AGU Fall Meeting Abstracts*, 2010.

[389] Y. Zhao, Z. Feng, Y. Zheng, and X. Cen. Existence of limit cycles and homoclinic bifurcation in a plant-herbivore model with toxin-determined functional response. *Journal of Differential Equations*, 258(8):2847–2872, 2015.

[390] T.N. Zimmerling and L.M. Zimmerling. A comparison of the effectiveness of predator odor and plant antifeedant in deterring small mammal feeding damage on lodgepole pine seedlings. *Journal of Chemical Ecology*, 22(11):2123–2132, 1996.

Index